HEAT PUMPS

HEAT PUMPS

J T McMullan and R Morgan

Energy Study Group
The New University of Ulster

Consultant Editor, N H Lipman

Adam Hilger Ltd, Bristol

McMullan, J. T.
  Heat pumps
  1. Heat pumps
  I. Title II. Morgan, R.
  621.402'5 TJ262

  ISBN 0-85274-419-6

Published by Adam Hilger Ltd,
Techno House, Redcliffe Way, Bristol BS1 6NX

The Adam Hilger book-publishing imprint is owned by
The Institute of Physics

Printed in Great Britain by Page Bros (Norwich) Ltd

D
621.4025
MACM

CONTENTS

PREFACE

The only reason why space heating is required to heat a building at all is that, due to the fact that the inside is at a higher temperature than the outside, it continually loses heat to the environment. This produces an environmental heat load which is equal to the heat demand of the building. One of the features of the heat pump is that it manages to reduce both this environmental burden and the energy requirement to supply the heat demand, which is itself unaffected. It achieves this by recycling some of the lost heat back into the building again. This is ultimately what happens whether the heat source is the outside air, the ground or the sea, and it emphasises the basic benefit that the heat pump has to offer—heat recovery.

Given that some 38% of national energy consumption is typically devoted to space and process heating, with only 8% being attributable to manufacturing, it is obvious where the largest potential for the application of heat pumps lies, at least in principle. Industry, however, is very sensitive to the importance of reducing manufacturing costs so that overt industrial heat recovery applications offer attractive possibilities for heat pump projects. Further, since industrial applications are frequently on a somewhat larger scale than domestic and commercial space heating installations, it is likely that more fuel is to be saved in any one plant.

Heat pumps, therefore, have an important contribution to make in helping us face the coming transitional period in energy use patterns. Indeed, since most of the alternative energy sources such as wind power, nuclear power, etc all favour the production of electricity, the heat pump becomes steadily more and more attractive as the transition progresses— assuming, of course, that it does progress.

One of the problems with heat pump development is that there is a tendency for engineers to believe that, since refrigeration has been available for at least 150 years, and since the heat pump is nothing but a refrigerator, there is no research needed or development work to be done. Nothing is further from the truth. Despite the similarities in the thermodynamic cycles, the heat pump is not a refrigerator. It is subject to much more rigorous operating conditions, to an exceptionally large range of load factors, to control problems that do not arise in refrigeration and to demands on component efficiency that have never been faced by the refrigeration industry.

In approaching the problem of writing this monograph, we decided that there were already sufficient books enthusiastically proclaiming the virtues of the heat pump at a fairly low level of scientific sophistication. Equally, there is a much smaller number which discuss the application possibilities in a way suitable for architects or building services engineers. There is a large gap, however, when one considers the more fundamental aspects of design and testing and this is the area at which the present work is aimed. We hope to present the new research worker in the heat pump field with enough background, and with a sufficient body of reference material, to enable him to avoid some of the pitfalls into which we have fallen. We have deliberately avoided a detailed discussion of application areas except for what could be regarded as some personal observations, contenting ourselves with an overview. The references in Chapter 5 give some more extensive reviews of this area.

The treatment assumes some knowledge of thermodynamics, electronics and the type of general science and engineering background that is required for work in this field. Fundamental research work on heat pumps tends to require a broad scientific education and a high tolerance of Freon. We have attempted in the references and bibliography to include material essential to the development of the subject matter and also useful background and ancillary reading. In this, we hope we have been sufficiently wide ranging to accomplish the task.

J T McMullan
R Morgan

1

INTRODUCTION

The refrigerator has today become such a ubiquitous item of domestic equipment that few of us ever stop to think about how it operates. It is only when confronted with a new name such as *heat pump* that we begin to think of the principles involved. At this stage we begin to appreciate that the refrigerator and the heat pump are one and the same thing, and that to a large measure the machine is called a refrigerator if our interest is primarily with the absorption of heat and the cooling of a chilled space; if we turn our attention to the heat rejected at the other end of the machine and to heating applications then we call it a heat pump; and if we are interested in the heat absorption, but this time in the cooling of an occupied space to comfort levels, we call it an air conditioner. Actually, these definitions are becoming a little blurred with the extension of the uses of the heat pump into industrial heat recovery applications and into drying, but they give a useful rule of thumb.

The heat pump/refrigerator/air conditioner is a machine which is capable of absorbing heat at a low temperature and rejecting it at a higher one. This smacks of wizardry since instinct tells us that it is forbidden by the Second Law of Thermodynamics. Our feeling is somewhat reinforced when we find that the heat rejected is greater than the energy we have supplied to operate the machine. This has introduced the 'something for nothing' effect which we also know to be impossible. All is not lost, however, as we soon realise that the mechanical work supplied to the machine was used to effect the transfer from the low temperature to the higher one, and that the energy rejected at the higher temperature includes both the mechanical energy and the thermal energy absorbed at the low temperature. The Second Law has survived another onslaught by the perpetual motion enthusiasts.

The concept of the heat pump is not new, and has been attributed generally to Lord Kelvin who produced a design in 1852. In fact, the vapour compression refrigerator designed by Jacob Perkin pre-dates this by some 18 years. Air-compression refrigerators were proposed by Oliver Evans of Philadelphia in 1805 and built by John Gorrie in 1849, while the first ice factory was apparently built in Australia by James Harrison about 1850.

(Harrison also first installed refrigeration plant in a brewery in 1851). The first man-made refrigeration produced by the evaporation of ether into a partial vacuum is generally credited to William Cullen of the University of Glasgow in 1748, but two earlier references to refrigeration technology are worth mentioning. In his translation of the early fifth century *Historiae adversum paganos* by Paulus Orosius, Alfred the Great (*c* 900) adds what appear to be verbatim reports of ninth-century voyages to the White Sea and the Baltic. In the second of these Wulfstan reports on the practice in Estonia of storing the dead indoors and above ground for several months before final cremation. The period of storage varied between one and six months depending on the wealth of the deceased, and that it was accomplished by freezing is clear from the closing sentences of the account: 'And there is among the Ests a tribe that can produce cold; and dead men lie there so long without decay because they bring the cold upon them. And if one set down two vats full of ale or water, they will arrange for both to be frozen over, whether it be summer or winter.'

The other example appears in Muirchu's seventh-century *Life of St. Patrick*. 'And, to make a beginning of the matter, he (the seer Lucetmael) put, while the others were looking, somewhat from his own vessel into Patrick's cup, to try what he would do. St Patrick, perceiving the kind of trial intended, blessed his cup in the sight of all; and lo, the liquor was turned into ice. And when he had turned the vessel upside down, that drop only fell out which the seer had put into it. And he blessed his cup again, and the liquor was restored to its nature; and all marvelled.

And after the trial of the cup, the seer said "Let us work miracles on this great plain". And Patrick answered and said, "What miracles?" And the seer said "Let us bring snow upon the earth." Then said Patrick "I do not wish to bring things that are contrary to the will of God." And the seer said "I shall bring it in the sight of all." Then he began his magical incantations, and brought down snow over the whole plain to the depth of a man's waist; and all saw it and marvelled.'

The heat pump *per se* did not appear, however, until Haldane built the first practical machine about 1930 and used it to heat his home in Scotland. He used the atmosphere as the heat source and backed it up with the local water supply when atmospheric conditions were not favourable.

There was little subsequent interest in the heat pump until the 1950s when some developments were made, but these were overtaken by the fall in the real price of oil and by some operating difficulties. With the expansion of the refrigeration industry, many of the problems with the earlier models were overcome and, with the rise in oil prices of 1973–4, there was a resurgence of interest once again. The main difficulty that the heat pump faces in breaking into the domestic heating market is one of initial cost. It is difficult to convince a home owner that he should invest in a capital intensive piece of heating equipment when he can acquire the same or greater heating power at half the price or less by buying an oil- or gas-fired boiler. At the moment, the running costs of a heat pump heating system are sufficiently smaller than those of an oil-fired system that there is a realistic pay-back period, but the same cannot be said when the competition is from natural gas. The running costs of the heat pump are comparable to those of a gas-fired system (at least in the United Kingdom) and there is no incentive to change, other than the fact that there is a net energy saving associated with the heat pump. At least part of heat pump research effort must be directed towards reducing the initial cost, while the rest must aim at improving the heat transferred per kilowatt of power supplied.

These two aims, while important everywhere, are not as critical in countries such as the United States where there is a cooling as well as a heating function to be performed. Here there is no other way of realistically providing the air conditioning function, and people are willing to pay for it. As a bonus, a heating system is provided which is at worst no more expensive to operate than the competition and the utilisation of the plant increases dramatically. This has the effect of reducing the cost of the heating system, though at the penalty of providing a less efficient heat pump for reasons that we will see later. Even so, this approach of finding a combination of a heating and a cooling load increases the effectiveness of the plant and improves the economies to be achieved. This factor will be discussed in more detail when we come to the applications of heat pumps in different areas. First, however, we must look at the principle of operation and at the different practical approaches to achieving a working unit.

1.1   Principle of operation

The heat pump and the heat engine can be shown schematically on the same diagram, as in figure 1.1. In the heat engine, heat is supplied at a high temperature T_h, mechanical work is performed and the residual unconverted heat is rejected at a low temperature T_c. If quantities of heat are labelled by Q and the mechanical work by W, then

$$Q_h = Q_c + W,$$

and the efficiency, η, of the engine can be taken as the ratio of the mechanical work extracted to the original heat energy supplied at the high temperature T_h. That is,

$$\eta = W/Q_h.$$

By contrast, in the case of the heat pump, the heat is supplied at a low temperature; mechanical work is supplied to effect its transfer to the high temperature, and the sum of the original heat pickup and the mechanical work is rejected to the high-temperature sink. Thus, the equation $Q_h = Q_c + W$ still applies, but the origins of Q_h and Q_c are different from those of the heat engine. The first complication arises when we try to define what we mean by the efficiency of the machine. Following the logic of the heat engine experience, we can define efficiency as the quantity of heat absorbed

3

from the cold source per unit of mechanical work required to transfer it to the sink, or alternatively, we could define it as the quantity of heat rejected at the high temperature per unit of mechanical work required. The first of these is used if our interest lies in the refrigeration capacity of the machine, as the amount of cooling per kilowatt is the important factor; the second is used if we are interested in the heating capacity of a heat pump. Even now, things are not straightforward. Since $Q_h = Q_c + W$, to define the efficiency in terms of the heat rejected per kilowatt of mechancial energy supplied means that the ratio Q_h/W must be greater than unity, which offends our ideas of the definition of an efficiency. Consequently, we change our nomenclature slightly to talk about a *coefficient of performance*, COP, defining separate coefficients of performance COP_h and COP_c to describe the heating and cooling performance of the machine. Hence, the coefficient of performance for heating is given by

$$COP_h = Q_h/W,$$

and the coefficient of performance for cooling by

$$COP_c = Q_c/W.$$

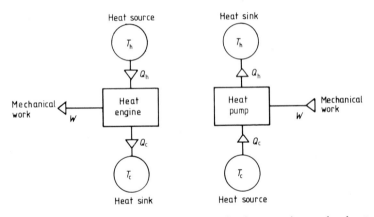

Figure 1.1 Schematic representations of a heat engine and a heat pump.

There is a temptation to take our earlier equation for the energy balance to suggest that

$$COP_h = COP_c + 1,$$

but this is only true in the ideal case. In a real system, because of heat losses, the *useful* heat rejected is less than the sum of the actual energy supplied and absorbed.

All of the heat pump cycles depend on some means of supplying an external energy source to provide the pumping action, and the most common is undoubtedly the *vapour-compression cycle*. At the moment it would appear that this is the only heat pump cycle which, because of the disadvantages in cost, size or efficiency of the other designs, is economically viable in the short term. This need not always be so.

Before looking at the vapour-compression cycle, it is perhaps worthwhile reminding ourselves of the ideal standard cycle, the *reversed Carnot cycle*, which is the refrigeration cycle analogous to the Carnot engine. This is well illustrated by using a temperature–entropy (T–s) diagram as shown in figure 1.2 and we can follow the sequence of operations on the working fluid (or *refrigerant*). From point 1 the fluid is compressed adiabatically to point 2, and thence is compressed isothermally to point 3. During this stage a quantity of heat Q_h is lost by the fluid. From point 3 the working fluid is allowed to expand adiabatically to point 4, and thence to expand isothermally back to point 1, completing the cycle and taking in a quantity of heat Q_c in the process.

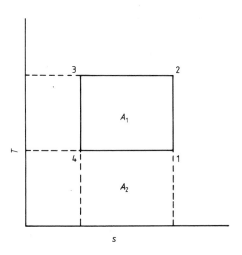

Figure 1.2   Temperature–entropy diagram for a reversed Carnot cycle
heat pump.

From the theory of the Carnot engine which is available in any standard thermodynamics text or, for example, McMullan *et al* (1976), we know that the efficiency is given by the work done per unit of heat taken in at the high temperature. That is,

$$\eta = W/Q_h = (Q_h - Q_c)/Q_h,$$

and this can be written in terms of the temperatures to give

$$\eta = (T_h - T_c)/T_h.$$

It can also be written in terms of the areas A_1 and A_2 on the T–s diagram as

$$\eta = A_1/(A_1 + A_2).$$

5

When we apply this to the heat pump coefficient of performance as defined earlier, we obtain

$$\text{COP}_c = Q_c/(Q_h - Q_c) = A_2/A_1 = T_c/(T_h - T_c),$$
$$\text{COP}_h = Q_h/(Q_h - Q_c) = (A_1 + A_2)/A_1 = T_h/(T_h - T_c).$$

These two equations are important as they represent the highest performance that can theoretically be achieved, though they are unobtainable in practice for a variety of reasons. The Carnot cycle is thermodynamically reversible, which carries with it certain limitations and restrictions. For example, the isothermal expansion and compression must be exactly that, and this is not possible to achieve in practice because any real heat exchanger requires a finite temperature difference between the working fluid and the heat source if it is to operate effectively. This leads to a departure from the ideal and causes the cycle to become irreversible. This in turn entails an increase in the entropy of the overall system and a corresponding increase in the energy input and decrease in the coefficient of performance. These factors will be returned to later, but their effect is to force a compromise to be reached between various financial, physical and other restrictions so that the actual machine is as close in operation as possible to the theoretical ideal. Even so, the expression for the coefficient of performance clearly indicates why the heat pump is of interest as an energy saving device. (This use of the term energy saving and its brother energy conservation is one of the less welcome developments to appear since 1973, but it has now been lexically absorbed and we have little option but to use it.) Consider the use of a heat pump to extract heat from the atmosphere at 278 K (5°C) and to deliver it to heat a house at 293 K (20°C). The theoretical coefficient of performance is 293/15 = 19.5, which is highly attractive since it implies that 19.5 kW of heat can be supplied to heat a house for the expenditure of only 1 kW. Even accepting that this is unrealistic by allowing 5 K temperature difference between the heat source and sink and the refrigerant, we get a COP_h of 298/25 = 11.9, and if we say that the other inefficiencies reduce the performance by about a factor of two, we end up with a COP_h of about 6, which is a very attractive prospect. In fact, this value is not achievable except under particular circumstances which are not common, but which we will return to later. Much more common values for the COP_h lie between 2 and 3, and while some manufacturers claim values significantly higher than 3 for domestic equipment these claims can usually be shown not to stand up to experimental confirmation. Improvement of the COP_h must be one of the main aims of heat pump researchers, though this improvement must not be achieved through an increase in the initial cost of the machines.

1.2 Vapour-compression cycle

The most common present route to producing a heat pump is the vapour-compression cycle. In this, the refrigerant is a condensable vapour, and the heat exchanger conditions are chosen so that liquid refrigerant is boiled in the low-temperature heat exchanger (evaporator) gaining latent heat of vaporisation, and is condensed in the high-temperature heat exchanger (condenser) so releasing the latent heat again, together with any heat that may have been added during the pumping process. Figure 1.3 gives a schematic diagram of the process, which is also shown in an ideal form in the pressure–enthalpy and temperature–entropy charts of figure 1.4.

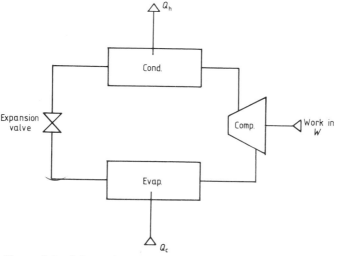

Figure 1.3   Schematic diagram of a vapour-compression cycle.

Starting at point 4, the low-pressure, low-temperature refrigerant is boiled in the evaporator, and emerges at point 1 in the form of a dry saturated vapour. From here, it is compressed adiabatically to a high-temperature, high-pressure superheated vapour at point 2. From point 2, the superheated vapour enters the condenser where it first of all loses the sensible heat associated with the superheating above the boiling point at the pressure of interest, and then condenses, in this case to the saturated liquid line at point 3. From point 3, the high-temperature, high-pressure liquid is expanded through an orifice to the much lower pressure and temperature of the evaporator. Some of the refrigerant flashes off and the refrigerant enters the evaporator as a low-quality vapour at point 4. The heat transfers are Q_c into the refrigerant from the heat source at the evaporator, and Q_h out of the refrigerant to the heat sink at the condenser. A quantity of heat is supplied through the work of compression between points 1 and 2, and the nature of the various steps is clearly shown in the P–h and T–s diagrams. It is assumed that the heat transfers at the heat exchangers occur through zero temperature difference, but the irreversible nature of the isenthalpic throttling process from 3 to 4 and the effect of the temperature difference through desuperheating are apparent. These show that

7

we are no longer trying to deal with an ideal process, and illustrate the first of the irreversibility losses that must be accounted for.

The performance of the cycle can be seen most easily on the $P–h$ diagram as enthalpies can be read directly off the axis and do not need to be calculated from areas and so forth, provided that kinetic and potential energy terms can be ignored. This is a very good approximation, and so the energy balance for the cycle can be written as follows:

1–2 work of compression	$h_2 - h_1$	(input)
2–3 heat of condensation	$h_2 - h_3$	(output)
3–4 isenthalpic expansion		
4–1 heat of evaporation	$h_1 - h_4$	(input).

Thus the coefficients of performances are

$$\mathrm{COP_h} = (h_2 - h_3)/(h_2 - h_1),$$
$$\mathrm{COP_c} = (h_1 - h_4)/(h_2 - h_1).$$

These two equations give us the basis for measuring the performance of any heat pump system, or at least for comparing the performance of different refrigerants, as the manufacturers of the refrigerants publish pressure–enthalpy–entropy–density data for each of the available materials. Naturally, there are deviations from these simple values, and many factors affect the performance. In practice, as well as the finite temperature difference effects already mentioned, there will be pressure losses in the pipework, valves, condenser and evaporator, and the compressor will not be 100% efficient.

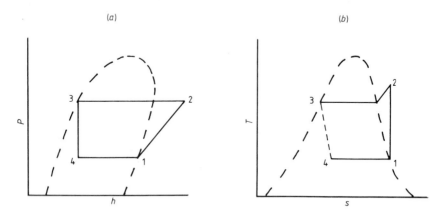

Figure 1.4 (*a*) Pressure–enthalpy and (*b*) temperature–entropy diagrams of a vapour-compression heat pump. Note that isenthalpic expansion (3–4) cannot adequately be represented on the $T–s$ chart.

Apart from these factors, the cycle is normally slightly different from the ideal one shown above for several reasons. First, the small reciprocating compressors appropriate to many heat pump applications have a very poor tolerance for liquid refrigerant. This is because the liquid, being incompressible, can cause extensive damage to valves and connecting rods if it is allowed to enter the cylinder of the compressor. For this reason, care must be taken to ensure that the refrigerant is fully vaporised (or nearly so) and the normal procedure is to slightly superheat the vapour in the evaporator. This is achieved by using a thermostatic expansion valve which controls the flow of refrigerant so that the evaporator is capable of superheating it by some preset amount, usually 5 K or so. Thus, the evaporator temperature depression is increased even more relative to the heat source temperature, and further, the efficiency of the evaporator is reduced because more of the surface is in contact with vaporised refrigerant with a correspondingly smaller heat transfer coefficient. At the other end of the machine, the output from the condenser is sometimes subcooled, which has the effect of increasing the heat rejected from the condenser and of increasing the amount of heat absorbed at the evaporator, as shown in figures 1.5 and 1.6. If this possibility is available, it provides a good way of increasing the performance of the heat pump, as more heat is pumped for the same compressor power. However, in most cases it is worth considering the amount of this gain in relation to the increase in condensing temperature that was necessary to allow the subcooling to occur in the first place. For refrigeration applications, the subcooling option may well be available, particularly in the winter, but for heat pump operations it is likely that the degree of subcooling should be minimised as shown in figure 1.7, where the effect of reducing the compressor work $h_2 - h_1$ is apparent if the condenser temperature is reduced so that there is no subcooling at the outlet. As will be seen later this is not a simple matter and the degree of advantage varies according to the operating regime and also with the refrigerant being used.

This is perhaps a useful juncture to discuss the properties required of

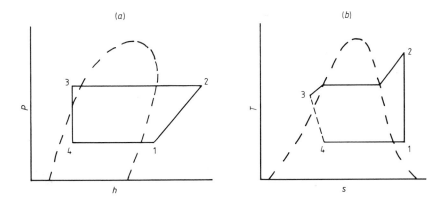

Figure 1.5 (*a*) Pressure–enthalpy and (*b*) temperature–entropy diagrams of a heat pump with superheating at the evaporator and subcooling at the condenser.

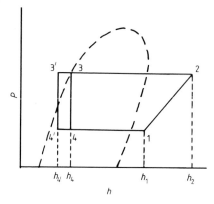

Figure 1.6 Comparison of heat absorption and rejection with and without subcooling of the condenser output.

any chosen refrigerant. From an examination of the $P–h$ diagrams, we can make the following observations. Firstly, the critical temperature of the refrigerant should be above the highest possible condensing temperature achieved in the cycle. It is also useful, though not essential, to have the minimum saturation pressure in the cycle above atmospheric pressure, as this ensures that any system leakage will be outward rather than inward. It is also beneficial if the saturated liquid line is steep (resulting from a low specific heat in the liquid phase) as this reduces the quality of the vapour at the inlet to the evaporator. In addition, the compression ratio between the vapour pressures at the evaporator and condenser temperatures should be as small as possible, and the condenser pressure should be reasonably low. Finally, the refrigerant should have the highest possible latent heat

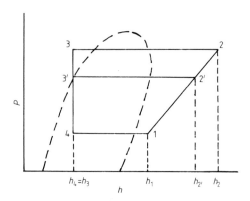

Figure 1.7 Effect of lowering the condensing pressure so that sub-cooling at the compressor outlet is eliminated. Compressor work is reduced from $h_2 - h_1$ to $h_{2'} - h_1$ and heat rejection is reduced to $h_{2'} - h_3$.

of evaporation, and the smallest possible specific volume of the vapour phase at the compressor inlet, as these two factors determine the volume of refrigerant that the compressor has to handle for a given heat output. As can be seen, these requirements are complex and in some cases conflicting. The situation becomes worse when we add the other constraints of chemical stability, low toxicity, non-inflammability, viscosity, cost and availability, and the result is that most installations are charged with one of a very small number of 'standard' refrigerants.

One other small digression that it is worth making here involves the subject of units. In common with other engineering areas, the refrigeration and heat pump industry is plagued with a diversity of units, with the American industry working in Imperial units (*sic*), and most of continental Europe working in (almost) SI units, that is metric quantities are used, but the unit of energy is the calorie (cal). The United Kingdom is caught between these two stools and has a tendency to work in either Imperial units or, increasingly, in SI, correctly using joule as the unit of energy. The one unit that lurks around the fringe and which is not readily understandable to the novice is the *ton of refrigeration*. A refrigerator is said to have a capacity of one ton if it can freeze one ton (US) of ice at 32°F in one day. That is, if it has a heat removal capacity of 12 000 Btu h^{-1} (almost exactly 3.5 kW). The origin of this unit is obvious, and it still appears in the literature and in conversation so it is worth a special mention. It is also worth noting that a one ton heat pump with a COP_h of 2.4 has a heat output of about 6 kW, which is a reasonable size for domestic applications.

1.3 Absorption cycle

The simplest absorption refrigerator consists merely of two flasks connected by a pipe with a valve (figure 1.8). If flask A contains the refrigerant (for example pure ammonia liquid) and flask B contains the absorbent (for example pure water), then if the valve V is opened, refrigerant vapour will flow from A to B since the refrigerant has a higher vapour pressure than the absorbent. The refrigerant vapour condenses in B, dissolving in the absorbent and releasing both the heat of condensation and any heat of solution, so raising the temperature of B and consequently the equilibrium pressure. In flask A the reverse process takes place. As the working fluid evaporates to replace the losses by mass transfer to flask B, both the temperature and the pressure fall until a new equilibrium pressure is reached for the whole system. The net result is a temperature difference between the two flasks with flask A at a lower temperature than flask B. If heat is applied to flask A, this newly established equilibrium is disturbed and mass transfer begins once more. As the working fluid condenses in B, heat is pumped.

This system is limited and will stop after a time. In order to regenerate it, the two fluids must be separated again and this can be achieved by heating flask B while at the same time removing heat from flask A. This

11

device has been used as an intermittent refrigerator since its invention by Sir John Leslie in 1810, using water as the working fluid and sulphuric acid as the absorbent.

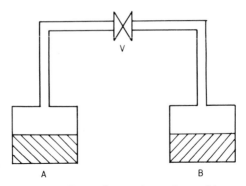

Figure 1.8 Intermittent absorption refrigerator.

In the evolution of this type of system to a continuous process, three factors have to be allowed for. Firstly, the absorbent and refrigerant have to be separated at high temperature and the refrigerant recondensed so that it can evaporate in the equivalent of flask A. Secondly, the absorbent has to be similarly cooled for the admixture of the vapour refrigerant in the equivalent of flask B. Thirdly, the dilute mixture of refrigerant and absorbent has to be continuously removed from the equivalent of flask B to the separator. This is shown in figure 1.9 where it can be seen that the absorption refrigeration system effectively allows the compressor of the vapour-compression system to be replaced by the combination of a liquid pump and a heat source which is different from that supplying heat to the evaporator. The evaporation and condensation stages of the vapour-compression units are retained with the evaporator becoming flask A, but the transfer of the vapour from the low to the high temperature is achieved by dissolving the cool refrigerant vapour in a non-volatile absorbent in the *absorber* (equivalent of flask B), and then pumping the solution to the higher pressure, as shown in figure 1.10. At the high pressure, the refrigerant is separated from the absorbent again by supplying heat to boil it off. It is then sent back to the condenser and thence back to the evaporator by way of an expansion valve. The pressure differential is produced by the heat supplied to the *generator* (the separator of figure 1.9) at the reva-porising stage, and physico-chemical processes, together with a very small amount of mechanical energy to pressurise the liquid solution, have been substituted for the much larger amount of energy required to compress a vapour. The contrasts and similarities between the two systems are shown in figure 1.11, where it can be seen that the absorption cycle also requires the introduction of a throttling device to ensure that the liquid is introduced to the *absorber* or mixing vessel at the same pressure as the refrigerant vapour from the evaporator.

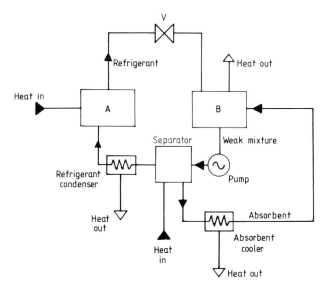

Figure 1.9 Modification of intermittent system to allow continuous operation.

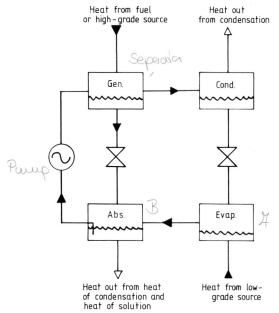

Figure 1.10 The absorption refrigerator. Refrigerant vapour from the *evaporator* is dissolved in the absorbent in the *absorber*, liberating heat of condensation and solution. The liquid mixture is pumped to the *generator* where heat is supplied from a fuel or other high-grade source to boil off the refrigerant once again. The hot vapour refrigerant is passed to the *condenser*, liberating its latent heat of vaporisation, while the absorbent is returned via a throttling device to the absorber. From the condenser, the refrigerant is returned via a throttling device to the evaporator, so completing the cycle.

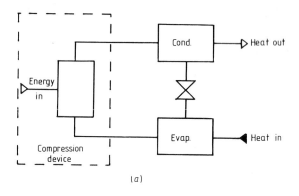

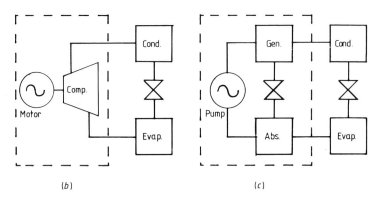

Figure 1.11 Contrast between vapour-compression and absorption machines: (*a*) generalised vapour–liquid refrigeration system; (*b*) vapour-compression cycle; and (*c*) absorption cycle.

From figures 1.9 and 1.10 we can see therefore that the absorption refrigerator is a heat-driven machine except for the small amount of power needed to supply the liquid solution pump. For this reason it is worth adding to our diagrams the complete system for the vapour-compression unit, that is including a steam turbine to supply the mechanical power, but excluding the electricity generation phase for simplicity. This is shown in figure 1.12, and the parallel with the absorption system is obvious. The particular advantage of doing this is that the vapour-compression system is now referred to a heat-driven system, and the COP can be formulated accordingly. This allows realistic comparisons on a primary fuel basis to be drawn between the performances of the two types of machine, though any such discussion must really include a consideration of the value of the fuel being consumed in the two cases. An absorption machine burning natural gas or gas-oil is burning an inherently more valuable fuel than an electrical vapour-compression machine running on electricity from a power station fuelled by residual oil or uranium. This distinction is not clearly drawn in present discussions of primary fuel efficiencies.

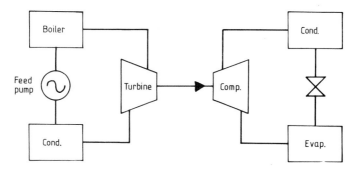

Figure 1.12 Vapour-compression cycle with heat source for compressor drive.

The efficiency of the heat engine can be written in terms of its source and sink temperatures, as can that of the heat pump. If we take the two sink temperatures to be the same, then the overall efficiency in heat terms can be written as

$$\text{COP}_h \text{ (primary energy)} = \frac{T_s (T_h - T_s),}{T_h (T_s - T_l)},$$

where T_h is the top temperature of the heat engine, T_s is the sink temperature (assumed common), and T_l is the refrigeration load temperature. This assumes Carnot efficiencies, and therefore represents the ideal. (Some ideal values of COP_h are shown in figure 1.13, which applies equally well to an absorption heat pump if T_h is replaced by $T_{\text{generator}}$, etc.) No practical machine will approach this figure. In practice, if we assume that the efficiency of electricity generation is $\frac{1}{3}$ and that the heat pump drive is an electric motor, then the primary energy coefficient of performance of the unit is $\frac{1}{3}\text{COP}_h$, and this is the figure that can be compared with the performance of an absorption machine. If the primary drive is not an electric motor, but an internal combustion engine, then the picture is complicated by the prospect of heat recovery from the engine itself, and this raises questions that will be discussed later. For the moment, let us return to the absorption machine.

The secret of designing a successful absorption heat pump lies in finding a solvent–solute pair which satisfies several criteria. For example the solvent should be non-volatile over the temperature and pressure ranges of interest, while the solute should be condensable from the vapour phase at the high end and also be capable of being vaporised at the low end of the range; the mutual solubility should be high; the refrigerant should not freeze at any conceivable evaporator temperature; the materials should be non-toxic, and so forth. The two pairs commonly used in refrigeration and air conditioning applications are lithium bromide–water, and water–ammonia. In the first of these, the absorbent is a concentrated lithium bromide solution and the refrigerant is water—obviously used for air conditioning

15

or heat recovery applications, and unsuitable for systems in which the evaporating temperature can fall below 273 K (0°C). In the second, water is the absorbent and ammonia is the refrigerant. Here there is an application to refrigeration and to sub-zero operation since the water never comes into contact with the sub-zero environment. Other solvent–solute pairs undoubtedly exist (for example E131–R22), but, by and large, they have not yet been adequately categorised.

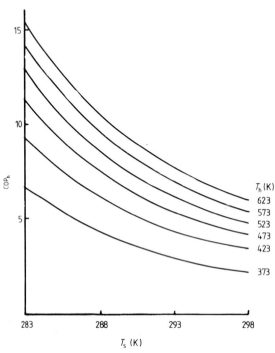

Figure 1.13 Ideal COP_h values for a range of heat engine source temperatures T_h and sink temperatures T_s for a given refrigeration load temperature T_l.

There are other differences imposed by the particular choice of refrigerant–solvent pair. For example lithium bromide is non-volatile and, as a consequence, the generator can be a simple still; whereas in the case of ammonia–water, fractional distillation equipment is necessary as the absorbent is volatile, and the refrigerant should be essentially free of absorbent. Too much absorbent in the refrigerant will hamper vaporisation in the evaporator.

In the absorbent circuit, on the other hand, there is normally a substantial amount of refrigerant; this is desirable for several reasons, but essentially it is to avoid excessive temperatures in the generator, and to ensure that a solid absorbent (such as lithium bromide) does not crystallise at any time.

The analysis of an absorption system is rather more complicated than that of the vapour-compression system. There are two separate fluid streams to be catered for and the mass flows must be in balance. Further, an understanding of the behaviour of the generator and absorber requires a consideration of the properties of fluid mixtures. This can be illustrated by considering the following examples of a lithium bromide–water machine, shown schematically in figure 1.14 where the cycle points of interest are numbered. It should be noted also that a new heat exchanger has been introduced between the input to the generator and the return absorbent flow. This is to reduce the amount of heat transferred back to the absorber, and is, in effect, to prevent the heat input to the generator from being passed to the cold part of the system. A First Law thermodynamic analysis (based on material balances, energy balances and equilibrium conditions) helps to gain an insight into system behaviour, though a Second Law analysis for a particular system based on the degradation of available energy (the energy available to perform useful mechanical work) will allow the identification of thermodynamic losses in the system and even possibly will point to their correction. However, let us confine our attention to a simple First Law analysis.

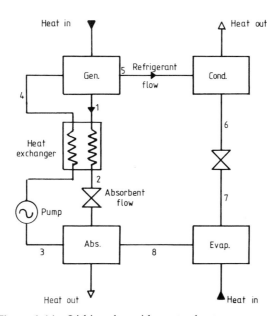

Figure 1.14   Lithium bromide–water heat pump system.

For a complete analysis, information will be needed regarding the enthalpy-concentration properties, the vapour pressures, the specific gravities, specific heats and viscosities of aqueous solutions of lithium bromide. These are to be found, for example, in ASHRAE (1974), and steam tables will provide the necessary data for pure water. A number of approximations

17

and assumptions are also made. Pressure drops, except at the expansion devices, are assumed to be negligible, while it is also assumed that the two phases are everywhere in equilibrium. In order to analyse a system, the temperatures of the four vessels must be specified, as must the heat extraction at the evaporator and the temperature drop ΔT across the liquid heat exchanger.

To proceed, table 1.1 is set up and a start is made on inserting the various values. Initially, the high and low pressures in the system (P_c and P_e) can be specified by the water vapour pressures at the known condenser and evaporator temperatures T_c and T_e. The enthalpies for water and steam at the same temperatures (h_6, h_7, h_8) are now also available from the steam tables, and the total mass flow per kg of refrigerant through the condenser and evaporator is, of course, unity. Thus points 6, 7 and 8 can be completely specified and point 5 partially so. The temperature at point 5 is determined by the generator conditions and can be assigned once these are specified (T_g). Following this, steam tables will again provide the enthalpy h_5 of the water vapour into the condenser. At this stage, the pressures are known at all points in the cycle, and the temperatures at points 1, 2, 3, 5, 6, 7 and 8. Because the generator solution is known to be boiling at temperature T_g and pressure P_c, the concentration c_1 and enthalpy h_1 at point 1 can be found from the enthalpy–concentration diagrams for aqueous lithium bromide solutions. Similarly, the known absorber temperature T_a and pressure P_e allow h_3 and c_3 to be specified. The fluid flow at the remaining points can now be determined from the material balances, which require that the mass flow into the generator w_4 must equal the mass flow out of it into the condenser and absorber ($w_1 + 1$). Thus, $w_1 c_1 = (w_1 + 1)c_4$ with $c_4 = c_3$, so that $w_1 = c_3/(c_5 - c_3)$, and $w_4 = w_1 + 1$.

The unknown quantities now remaining are h_2, h_4 and T_4. h_2 can be determined from the enthalpy–concentration diagram, since T_2 etc are known, and h_4 can be calculated from an energy balance:

$$h_4 = h_3 + (h_1 - h_2)w_1.$$

Subsequently, T_4 can be determined from the enthalpy-concentration diagrams.

In this way, the complete cycle can be specified, taking as the basis of calculation a refrigerant circulation of 1 mass unit. In order to calculate the output of a real system, the enthalpies and flow rates have to be scaled up to produce the required heat extraction. Thus, if the heat extraction at the evaporator is q_e per unit time, the mass flow of refrigerant will be

$$w_5 = q_e/(h_8 - h_7),$$

and the last two columns of table 1.1 must be scaled up by this factor. Thus,

| heat extraction at the evaporator | $= q_e$ |
| heat rejected at the condenser q_c | $= w_5 (h_5 - h_6)$ |

Table 1.1 Conditions in the lithium bromide cycle. (Reference points are identified in figure 1.14.)

Point	Temperature†	Pressure	Concentration of LiBr	Fluid flow/kg of refrigerant (kg)	Enthalpy (kCal/kg)	
1	T_g	P_c	c_1	w_1	h_1	
2	$T_g - \Delta T$	P_c	c_1	w_1	h_2	
3	T_a	P_e	c_3	$w_1 + 1$	h_3	
4	T_4	P_c	c_3	$w_1 + 1$	h_4	
5	T_g	P_c	0.0	1.0	h_5	⎫
6	T_c	P_c	0.0	1.0	h_6	from steam
7	T_e	P_e	0.0	1.0	h_7	tables
8	T_e	P_e	0.0	1.0	h_8	⎭

† T_c, T_e, T_g are defined by the cycle specification.

19

heat rejected at the absorber q_a	$= w_5(h_8 + w_1 h_2 - w_4 h_3)$
heat supplied at the generator q_g	$= w_5(h_5 + w_1 h_1 - w_4 h_4)$
heat transferred by the heat exchanger	$= w_5[w_1(h_1 - h_2)]$

The total heat rejected by the heat pump is $q_c + q_a$, while the heat supplied is q_g (plus a small amount of electrical energy to drive the solution pump). The coefficient of performance for heating is therefore given by $\text{COP}_h = (q_a + q_c)/q_g$. Typically, this will have a value in the range 1.3–1.5.

This simple analysis has assumed that the lithium bromide concentrations are those appropriate to saturation at the temperatures given. In practice, this would not be the case since such operation would run the risk of crystallisation of the salt. In real equipment, therefore, lower concentrations are used to ensure the presence of liquid absorbent at all points of the circuit. Another factor concerns the selection of the temperatures of the various vessels. These are interrelated to a large extent and depend on the application. Since heat can be extracted from both the condenser and the absorber, it is sensible to set the condenser temperature higher than the absorber temperature, which should be below 40°C (313 K) to reduce the danger of crystallisation. The generator temperature is related to the condenser temperature and is chosen in such a way that the absorbent concentration is within the desired range over all operating conditions of the system.

The analysis of an ammonia–water system is similiar to that given above, except that the system is necessarily more complicated, partly because of the fractional distillation mentioned earlier, and partly because the high generator temperatures needed make it desirable to transfer some heat by means of a coil in the stripping section. Also heat is usually removed in a partial condenser which generates reflux for the system. For more details of the type of calculations required, the reader is referred to ASHRAE (1974) where a sample calculation is given for an ammonia–water refrigeration plant.

1.4 Gas cycle

This cycle is based on using some gas—frequently air—as the working fluid, and it might be thought that the reversed Carnot cycle might be appropriate. Unfortunately, however, the adiabatic stroke requires a high stroke speed, while the isothermal stroke requires a low speed. Such a variation in stroke speed is not practicable in a real machine and instead, successful gas cycle refrigerators and air conditioners operate on the reversed Joule cycle. Usually, air is the working fluid.

The major design difference between the vapour-compression and reversed Joule cycle machines is that the latter employs an expansion device instead of a throttling valve. This is necessary because if a gas is passed through a throttling valve the temperature drop is small. Indeed, if the gas is perfect, there is no temperature drop at all. If a significant temperature drop is to be achieved, then the gas must be made to do work and so suffer a reduction in internal energy.

The reversed Joule cycle is shown in figures 1.15 and 1.16. Low-pressure air is taken in by the compressor at point 1 and compressed adiabatically (ideally) to a high-pressure, high-temperature state where it passes through a heat exchanger giving up some of its heat. It is now allowed to expand through an expansion device doing useful work which may, for example, be fed back to the compressor. The air cools during this expansion and then finally picks up heat again in the low-pressure heat exchanger before returning to the compressor to complete the cycle. In an ideal cycle, the compression and expansion are adiabatic and the heat transfers take place at constant pressure. Obviously this is not achieved in practice.

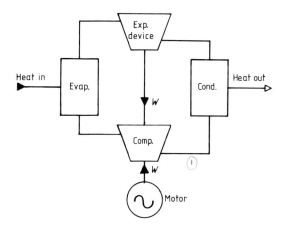

Figure 1.15   Reversed Joule cycle.

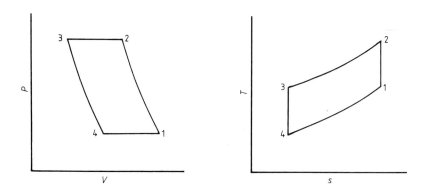

Figure 1.16   Pressure–volume and temperature–entropy diagrams for the reversed Joule cycle.

Analysis of the system is straightforward using the T–s diagram. Since the heat exchanges occur at constant pressure, the entropy changes can be written as

$$s_2 - s_3 = s_1 - s_4 = C_p \ln (T_2/T_3) = C_p \ln (T_1/T_4),$$

provided that the specific heat C_p can be taken as constant over the temperature range of interest. Thus T_4 is given by

$$T_4 = T_1 T_3/T_2.$$

The coefficient of performance is given by the ratio of the heat delivered $C_p(T_2 - T_3)$ to the work done. The work done, in its turn, is given by the difference between the compressor work $C_p(T_2 - T_1)$ and the work done by the expanding gas $C_p(T_3 - T_4)$. Thus the coefficient of performance (for heating) is

$$\text{COP}_h = \frac{T_2 - T_3}{(T_2 - T_1) - (T_3 - T_4)} = \left(1 - \frac{T_1 - T_4}{T_2 - T_3}\right)^{-1}.$$

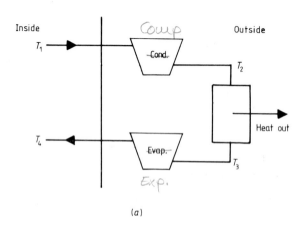

(a)

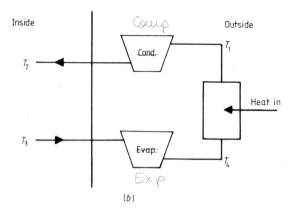

(b)

Figure 1.17 (a) Air conditioner and (b) heat pump modes of operation for an air-cycle refrigeration system.

Hence, from the earlier relation for T_4,

$$\text{COP}_h = (1 - T_1/T_2)^{-1}.$$

This result can be related to the compression ratio if we continue the assumption that the compression and expansion are adiabatic, whence $P^{1-\gamma}T = \text{constant}$. If R denotes the pressure between the heat exchangers then

$$T_3/T_4 = T_2/T_1 = R^{(\gamma-1)/\gamma}$$

and the coefficient of performance becomes

$$\text{COP}_h = (1 - R^{(1-\gamma)/\gamma})^{-1}.$$

The value of γ for air is approximately 1.4.

The advantage of the gas cycle using air as the working fluid is that one of the heat exchangers can be dispensed with. Thus, in an air conditioner (figure 1.17a), the 'evaporator' is omitted; air from the space to be cooled is drawn directly into the compressor, compressed, has heat removed while it is at the high temperature and then is expanded back to ambient pressure to emerge at a lower temperature, directly into the conditioned space. In a heat pump (figure 1.17b), the 'condenser' is removed; air is drawn from the space to be heated, expanded to a low pressure at which it can take in air from the outside heat source and is then recompressed to ambient pressure, being discharged at a higher temperature directly back to the space to be heated.

Thus, in principle at least, the reversed air cycle provides a light and inexpensive possible route to heat pump development.

1.5 Thermoelectric heat pumps

The Peltier effect, discovered in 1834, provides another potential route to heat pump development. It depends on the fact that if an electric current is passed round a circuit made of two dissimilar materials, one junction between the two materials will be heated and the other cooled. That is, heat is moved from one junction to the other. This is the opposite of the Seebeck effect in which an applied temperature difference produces an electrical potential difference. Successful operation depends on the discovery of materials which will provide a high thermoelectric effect, and combine it with low thermal conductivity (so that heat will not travel straight back along the connecting wire) and high electrical conductivity (so that ohmic heating is minimised). Until the advent of semiconducting materials, this combination of properties could not be found, but alloys of bismuth, antimony, selenium and tellurium all show hopeful characteristics.

Since no single physical property of the material is the determinant of performance for heat pumping applications, a figure of merit Z has to be defined which combines the Seebeck coefficient S with the electrical resistivity ρ and thermal conductivity k:

$$Z = S^2/4k\rho.$$

Typical values of Z can be as high as $0.003\ K^{-1}$ but this value is not sufficiently high to make thermoelectric heat pump performance competitive with existing vapour-compression units. A value of about $0.006\ K^{-1}$ would be required (Farrel 1975).

Because of this, and the problem of cost, it seems unlikely that thermoelectric heat pumps will offer a challenge in the near future, except in certain specialised areas of application for which cooling is probably the dominant function.

1.6 Sources of heat

A heat pump is at its most efficient when the source of heat and the sink are, as nearly as possible, at the same temperature. Consequently, it is worthwhile looking for the warmest possible heat source available. It is also worth remembering that the heat will be available at a temperature less than that needed for its direct utilisation, as otherwise there would be no necessity for the heat pump at all. Commonly used sources are air, water (which can be either from a stream or possibly could be ground water in suitable sites), and the ground. These are examples of natural sources and can be supplemented by industrial waste water, or exhaust air from mechanical ventilation systems, and so forth. Of these, the atmosphere is undoubtedly the most commonly used source.

In an air source heat pump, outside air is fanned over the evaporator coils, which are usually finned to increase the air-side heat transfer area, and a temperature difference of 5–10 K is typically established between the air stream entering the evaporator block and the refrigerant inside the coils. Air is an eminently suitable heat source because it is continuously available at all possible locations and leads to the use of equipment that is of reasonable size and which has relatively low operating and installation costs. Further, the design of air source heat pumps lends itself to factory production of standard units.

The disadvantage of air as a heat source is the fact that it varies rapidly in temperature and, particularly, the heat source temperature is low in winter when the heat demand is greatest. Further, the atmosphere contains large quantities of water vapour, and at low air temperatures, this will condense out onto the evaporator coil and freeze. This frost build-up has to be removed periodically to ensure that the coil block does not steadily accumulate more and more ice until it begins to resemble an advertisement for toothpaste. This is the much publicised 'defrost problem' (see Chapter 4), and it represents an important requirement for every air source heat pump design.

The fact that the air temperature is lowest at the time when the heat demand is greatest has other consequences as well. The coefficient of performance of the heat pump will be lower at low air temperatures (for the same condensing temperature) and so the heat output of the unit will be less than it would be at other times of the year. Thus, the unit has to be sized to meet the total demand at some design temperature, and at all temperatures greater than this the normal part load problem of any heating system is compounded by the increased capability of the heat pump to produce heat. The effect is obvious in principle from the expression for the efficiency of a Carnot heat pump $[T_h/(T_h - T_c)]$, and would suggest a rise in maximum theoretical COP_h from about 6.5 when the air temperature is 273 K (0°C) (corresponding to an evaporating temperature of about 268 K (−5°C)), to about 7.3 when the air temperature rises to 283 K (10°C). This is bad enough, but additionally, the effect is compounded by the way in which the components of any practical system react to changing temperatures. For example, the refrigerant vapour density at the inlet port of the compressor decreases as the evaporating temperature falls so that the same volumetric displacement represents a lower mass flow and therefore a lower input power to the compressor. This combines with the reduced COP to result in even less heat being rejected than the COP variation would suggest. This is a subject to which we will return in Chapter 2, but the nett result is that typically a unit which is designed to produce 9 kW of heat at an air source temperature of 273 K may well produce 12 kW at an air temperature of 283 K. At the same time, the heat demand of the house will have dropped from 9 kW to 4.5 kW (approximately) so that, instead of operating at a load factor of 50% as would a conventional heating system, our heat pump is operating at a load factor of only 4.5/12, or about 37%. This introduces the desirability of examining ways of modulating the capacity of heat pump systems to better match the conditions under which they are operating (see Chapter 3).

Water is another widely used source of heat. In West Germany and Denmark there is a large programme for developing domestic heat pump systems using ground water as the heat source. The water may be pumped out of the ground and through the evaporator, or it may be tapped by burying the evaporator coils in the ground and removing the heat *in situ*. The main advantage of this ground water as a heat source is that it is at an almost uniform temperature throughout the year. Thus, the problems introduced by the variations in temperature with air source machines are removed, and the load factor pattern reflects more closely that of a conventional heating system.

In addition, the problems of defrosting are eliminated, and as the heat transfer from the water to the refrigerant is better than for air, it is likely that evaporator design can be made more efficient—and certainly more compact.

On the debit side, however, it is undoubtedly true that the site work is more expensive for a ground water source heat pump than for an air source machine, and it is also less likely that the units can be readily factory-assembled in packaged units which can be taken to the site and easily installed. Most important, however, is the fact that access to suitable ground water sources is much more difficult than to a suitable air supply so that the machines are not so universally applicable.

The same criticism also applies to any other water source and so applies to all water source heat pumps to a greater or less extent. In addition, there is a larger seasonal temperature variation for surface water of any kind, so that river water or towns water is inherently less satisfactory than ground water, but is better than air; that is, if it is available in a given site.

The use of the ground as a heat source is more expensive again as it requires extensive heat exchange coils buried in the ground. These coils may contain the refrigerant itself, or preferably (because of the cost and the undesirability of having pressurised refrigerant piping buried underground and subject to possible damage during subsidence or site work at a later date) they may contain a brine solution which then passes through the heat pump evaporator. Conventionally, the ground coils are laid out in trenches at a depth of about 0.5–1.0 m, but some work has been done on an alternative system where they are grouped together in a vertical column which is sunk into the ground to a depth of several metres. This is cheaper and requires less actual ground area. Ground coil heat pumps are basically using the solar energy stored in the ground, so that over the heating season the ground becomes progressively cooler as heat is removed. During the summer, the heat deficit is restored by the sun and it is essential to ensure that the ground coil area is sufficiently large to ensure that this recovery can occur. Once again, the main advantage of the ground coil system over the air source machine is the constancy of the supply temperature, even though it reduces steadily throughout the heating season, but expense and the difficulties of installation in situations where access to fairly large ground areas is possible probably ensure that the air source machines will remain the most common heat pump systems in the coming years.

If access to a higher temperature source of low-grade heat is possible, then much better coefficients of performance are achievable, so that industrial effluent and power station cooling water, for example, become obvious sources of heat for domestic and commercial heat pump systems. Once again, the problem is the availability of both suitable heat sources and heat loads at convenient distances from each other so that the combination is possible. Much more interesting is the use of heat pumps for industrial and commercial heat recovery, where the low-grade waste heat being rejected can be further degraded (with environmental advantages) while a significant proportion of it is upgraded to a useable temperature and is returned to the factory or offices. This is an application of great potential importance and we will return to it in Chapter 5.

Meanwhile, however, we shall continue by looking at the components of heat pumps themselves and at system behaviour in rather more detail. In the ensuing chapters we will concentrate on the vapour-compression heat pump as it is the most developed and it is the one most likely to make an important contribution in the short term.

2

EQUIPMENT, REFRIGERANTS AND SYSTEMS

2.1 Equipment

Any heat pump has three basic components: the heat exchangers, the compressor and the expansion valve. Certainly the details may change in that the compressor is replaced by a solution pump and heat source in the absorption heat pump, or the expansion valve may be replaced by a turbine in the air cycle machine (or in some types of proposed heat driven heat pumps to which we will return briefly later). However, in general terms, there are only three component classes. We will concentrate our attention on the vapour-compression heat pump and its components, once again using the present state of the art and probable short term applicability as our justifications.

2.1.1 *Heat Exchangers*

Good heat exchangers are essential to the optimum performance of any heat pump system, and there are basically two types in common use. For air-refrigerant heat transfer, finned tube heat exchangers are most common, while for liquid-refrigerant heat transfer the shell and tube type is more common. These are illustrated in figures 2.1 and 2.2. In view of the fact that the refrigerant vapour being discharged from the compressor is superheated to a greater or lesser extent, there is a case for using a counterflow heat exchanger in the condenser and, if used, these normally take the form of a tube-in-tube design. Thus, two condenser options are a small tube-in-tube desuperheater together with a conventional shell-and-tube condenser (which provides its own liquid receiver because the refrigerant is in the shell), or a larger tube-in-tube countercurrent heat exchanger which acts as both desuperheater and condenser. This type requires the inclusion of a separate liquid receiver as there is no useful storage volume within the heat exchanger itself. This provision for a liquid receiver is an essential feature of any heat pump as the behaviour of the system varies according to the operating conditions, with the mass flow rate changing etc, so that storage space must be provided to stabilise the supply of liquid refrigerant to the evaporator. Further, if for any reason the system has to be opened (for example for compressor maintenance), the entire refrigerant charge will be lost unless it is possible to compress it into a storage vessel which can be valved off.

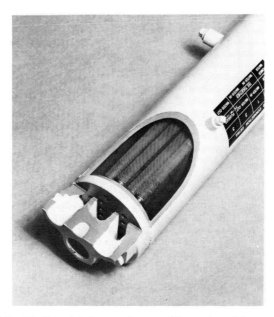

Figure 2.1 Shell and tube condenser. (Reproduced by courtesy of Dunham-Bush Ltd.)

There have also been proposals to use flat plate extended surface heat exchangers as evaporators for air source heat pumps, but so far these are limited by the sheer physical size involved in obtaining the same total heat transfer rate as can be achieved by the more conventional finned tube coil blocks.

Accurate sizing of heat exchangers for the particular heat pump installation is important. Unfortunately, the design of heat exchangers is complex and the heat transfer regimes are very complicated, so that accurate calculations are next to impossible for practical cases. Nonetheless, for research purposes it is important to be able to predict accurately how a heat exchanger will perform. At the condenser, the complications arise through the complex heat transfer processes on the refrigerant side (where the refrigerant will exist as superheated vapour, condensing vapour and subcooled liquid), and also on the heat sink side, particularly in the case of an air-cooled condenser where the humidity of the air has its part to play. These pale into insignificance, however, when compared with the difficulties of handling the analysis of an air source evaporator. Here there are boiling and superheating to account for on the refrigerant side and in addition, as will be seen later, there is the difficulty of allowing for the influence of the lubricating oil on evaporator performance. Some refrigerant is always dissolved in the lubricating oil, and this causes difficulties in predicting the evolution of the evaporation process; in effect, the published refrigerant data are no longer truly valid. On the air side of the heat exchange surface, the processes involved additionally include condensation of water vapour from the air and possibly its freezing, as well as the 'normal' processes of sensible heat transfer from the air and water vapour.

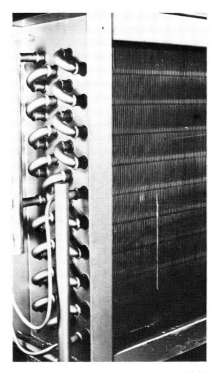

Figure 2.2   Air source evaporator coil block.

The influence of moisture in the air is an important factor because, in the UK climate, up to 15% of the heat collected at the evaporator may be arising through the latent heat of condensation of water vapour (McMullan *et al* 1980). The treatment of this problem is complex and not very well documented. Most existing theoretical work has been concerned basically with heat transfer in the case of a small amount of non-condensable gas entrained with a condensable heat exchange fluid. The case involved here is that of a small amount of condensable vapour entrained with a non-condensable heat transfer medium. There has been a modicum of extension of existing models (Hiller and Glicksman 1976, Hodgett and Lincoln 1978) to what are essentially dehumidification applications, but the agreement between theoretical prediction and experiment is often poor (see figure 2.3) and the theoretical models usually break down in certain humidity regions—unfortunately those of interest to heat pump modellers. One typical fault is the prediction under certain circumstances of negative moisture removal rates—not a common phenomenon with most evaporators.

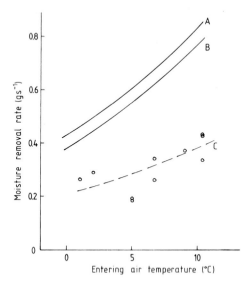

Figure 2.3 Moisture removal performance of air source evaporator. Relative humidity = 85%, $\Delta t = 15°C$. A, Colburn–Hougen model; B, McElgin–Wiley model; C, experimental data.

While detailed modelling of evaporator performance remains a difficult and sometimes a seemingly intractable problem, practical design criteria are met by the much simpler expression

$$q = UA\Delta T,$$

where q is the heat flow, A is the effective heat transfer area, ΔT is the average effective temperature difference and U is a heat transfer coefficient which is a complex function of the materials involved and the conditions under which the heat transfer is taking place. ASHRAE (1974) gives a range of equations for determining U.

This equation shows that the ways of increasing the amount of heat transferred are either to increase the temperature difference, or to increase the effective heat exchange surface area. For air source coils, this can be done by using fans to move the air more rapidly over the surface, so increasing the size of the effective surface. By and large, sufficient information on heat exchanger performance for most practical purposes can be obtained from the manufacturers.

2.1.2 *Compressors*

The compressor can be one of four basic types: reciprocating, rotary sliding vane, centrifugal, or rotating screw. Some differences are evident from figures 2.4, 2.5, and 2.6, and are emphasised by their range of application. The last two types are almost wholly used for large-capacity machines, while the first two, and primarily the reciprocating types, are used for smaller application areas. This generalisation is rather sweeping as there *are* large-capacity reciprocating compressors, but it does provide a rough rule of thumb guide. Because of lubrication and other problems, there are also some implications for the choice of compressor dictated by the temperature regime chosen for a particular application. Since the largest *potential* impact of heat pumps on the UK energy picture lies in their adoption in domestic space heating systems, we will concentrate primarily on the smaller compressor types, and in particular on reciprocating compressors. These are available in three types: hermetic, semihermetic and open. In the hermetic type (figure 2.7), the compressor and its motor are contained inside a hermetically sealed outer case. Cooling of the electric motor is either by radiation and convection of heat from the case, or is achieved by drawing the cold refrigerant from the evaporator over the motor on its way to the compressor (suction cooling). Usually, with hermetic

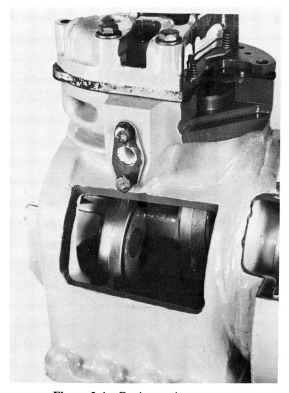

Figure 2.4 Reciprocating compressor.

Figure 2.5 Rotary sliding vane compressor.

compressors, a considerable effort has been made towards eliminating noise, by mounting the motor-compressor assembly on springs within the casing and so forth, and, because of the hermetic type of construction, the risk of oil or refrigerant leaks is small. These compressors are commonly seen in domestic refrigerator and deep freeze units; they are very reliable,

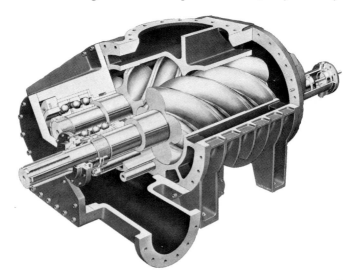

Figure 2.6 Rotating screw compressor, swept volume 5750 m³ h⁻¹, 2950 rpm. (Reproduced by courtesy of Stal Laval Ltd.)

Figure 2.7 Hermetic compressor.

silent in operation and, though they do exist in capacities up to about 5 kW for specialist applications, the range normally ends at about 1 kW. Hermetic compressors must be replaced in the event of compressor failure.

Semihermetic compressors are the units commonly seen in commercial and industrial installations. They consist of a compressor block which includes the casing holding the electric motor (figure 2.8). Once assembled, end plates are bolted over the block, making the assembly refrigerant and oil tight. Cooling can be either by radiative and convective heat loss from the case, or once again, by suction cooling. This type of unit is noiser than the hermetic compressors, primarily because there is no noise suppression between the compressor and motor and the casing, but the unit is amenable to disassembly for repair and maintenance.

The open compressor represents a further disassociation of the compressor itself from the drive mechanism. It consists simply of the compressor, with the crankshaft passing through shaft seals in the casing, and with the separate motor being coupled either directly or by belts. This type of compressor has the flexibility to allow any suitable type of prime mover to be adopted, but has the disadvantage that the pressurised refrigerant system is not completely enclosed and is potentially subject to leakage through the shaft seals. Obviously, however, it is immune to at least one type of failure—the *burn-out*, in which acids created by chemical reactions

33

between the refrigerant and residual water vapour in the system can attack the varnish on the motor windings, creating more acid and so forth until eventually the motor burns itself out through a short circuit in the windings. In a properly commissioned system this is a very small risk, but it can happen and the consequences are serious; the compressor and motor must be replaced, and the system must be completely cleaned to prevent a reoccurence. Obviously, if the motor is completely isolated from the refrigerant, burn-out cannot happen.

Figure 2.8   Semihermetic compressor.

The choice of compressor depends partly on application and partly on design philosophy and price. Hermetic compressors are cheaper than their semihermetic counterparts which are, in turn, cheaper than open compressors, and the price structure fairly closely reflects the numbers of each type that are used throughout the world.

2.1.3   *Expansion devices*

The remaining essential component is the device which reduces the refrigerant pressure and temperature so that it will pick up heat from the low-temperature source. This device can be a type that allows the expanding refrigerant to perform useful work (required for an air cycle unit) or, much more commonly, it can be a straightforward expansion device—essentially a pin-hole orifice. In small-scale equipment such as domestic refrigerators, it is usually a length of capillary tubing, the length and diameter of the capillary being predetermined by the most likely conditions that the unit is expected to encounter. In larger plant, or in systems where the

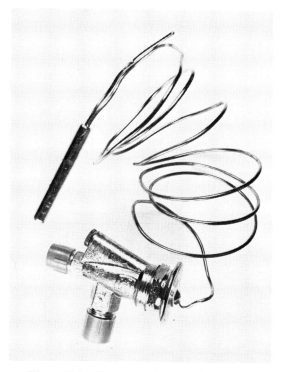

Figure 2.9 Thermostatic expansion valve.

range of control is likely to vary over too wide a range for a capillary tube to be effective, an *expansion valve* is used. If the conditions are relatively static, this valve may be one with a fixed orifice size, or with a regulated pressure drop, but in most applications the *thermostatic expansion valve* is used (figure 2.9). The thermostatic expansion valve serves a very important dual purpose. Not only does it reduce the refrigerant pressure and temperature, but it also regulates the flow of refrigerant so that the evaporator receives just the amount that can be evaporated. This is achieved as shown in figure 2.10. The orifice of the expansion valve is restricted by a needle. This needle is attached to a diaphragm which is exposed on one side to the refrigerant in the system, and on the other side to a vapour contained in the bulb B. If the pressure in the bulb B is higher than that in the refrigerant system, the diaphragm will move downwards, opening the needle valve, and allowing more refrigerant through. If the pressure in B is lower than that in the system, the needle valve is closed, restricting the flow of refrigerant. Additionally, an adjustable spring is included to ensure that the pressure in B must be somewhat higher than that in the system before the balance point of the valve is reached. Usually, the bulb B is filled with the same refrigerant as that in the system, but there may be other additives to improve the control characteristics of the valve.

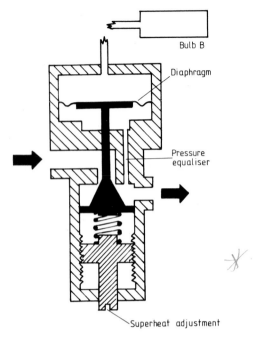

Figure 2.10 Thermostatic expansion valve operation.

The control system operates as follows. The bulb B is attached to the suction pipe of the compressor as shown in figure 2.11, making sure that there is a good thermal contact. If too much liquid refrigerant is admitted to the evaporator, so that it cannot all be evaporated, then liquid will appear in the suction pipe, and B will be at the boiling temperature of the refrigerant in the sytem. As a consequence, the vapour pressure in B will be the same as the refrigerant pressure in the system and there will be no

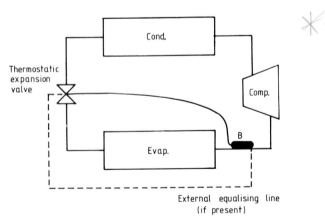

Figure 2.11 Positioning of thermostatic expansion valve in system.

net force on the diaphragm due to the vapour pressure balance. The adjustment spring, however, is set to add to the pressure on the refrigerant side of the diaphragm, and so the diaphragm moves up, closing the needle valve, and reducing the supply of refrigerant. At this reduced refrigerant supply, the evaporator is more capable of fully evaporating its intake, and the process continues until a stage is reached where the evaporator fully evaporates the refrigerant before it reaches the end of the coil block. If this happens, the latter part of the coil block transfers sensible heat from the heat source to the refrigerant, superheating it so that it emerges from the evaporator at the bulb B with a temperature greater than the saturation temperature appropriate to the evaporator pressure. As a consequence, the bulb B is at this higher temperature and the refrigerant charge inside it has a higher equilibrium pressure, which is transmitted to the upper side of the diaphragm in the valve. If this pressure is greater than the combined effects of the refrigerant system pressure and the adjustment spring, the diaphragm will move downward, opening the needle valve and allowing more refrigerant through.

In this way, the thermostatic expansion valve controls the flow of the refrigerant to be that which the evaporator is just capable of evaporating under the prevailing conditions. The spring ensures that the refrigerant flow from the evaporator to the compressor is fully evaporated and, indeed, slightly superheated, but not so much that a large fraction of the evaporator is used for transferring sensible heat to the refrigerant vapour. Thus, the valve is acting to optimise the utilisation of the evaporator. The small degree of superheating (usually about 5 K) is important as the rise in temperature is the only simple way of ensuring that vaporised refrigerant and not liquid is being admitted to the compressor.

If the evaporator is physically large so that there is a noticeable pressure drop from inlet to outlet, then the refrigerant-side chamber of the expansion valve diaphragm is usually connected to the outlet side of the evaporator. This ensures that the diaphragm sees the 'correct' saturated vapour pressure at the location of the bulb B. This type of valve is said to be *externally equalised*.

One other factor is important here for the smooth operation of the system. If the maximum orifice size of the valve is too small under certain operating conditions, then insufficient refrigerant will be admitted to the evaporator to utilise it fully. In addition, because the orifice is too small, the evaporator pressure will be reduced—producing lower temperatures, higher superheat levels and increased thermodynamic losses. If the maximum orifice size is too large, however, then a condition can be reached in which the valve, sensing that the output from the evaporator is excessively superheated, opens and allows an overly large amount of liquid refrigerant to enter the evaporator. The fact that the evaporator cannot cope with this quantity will remain undetected until a slug of liquid arrives at the bulb B, which will then close the valve abruptly and reduce the refrigerant flow until it again senses superheating of the evaporator output. This effect is called *hunting*, and with a correctly sized valve it is eliminated by increasing the tension in the adjustment spring, so changing the sensitivity of the valve to match the system to which it is connected. Hunting cannot happen with a valve that is too small, and cannot be eliminated with a valve that is excessively oversized.

A variant of the thermostatic valve, with an electrically operated control capsule, is sometimes used instead of the vapour pressure controlled valve. In its usual form it is controlled by a thermistor which senses liquid level at the exit from the evaporator. If liquid refrigerant appears, the thermistor is cooled more vigorously, and the change of resistance is used to close the valve orifice somewhat, restricting the flow of refrigerant. The main advantage of the electrical valve is that it is independent of the refrigerant in the system. This can be turned to advantage in heat pumps, especially in experimental situations, because it allows the use of unusual refrigerants without the necessity of ordering specially charged valves. A more important advantage is the possibility of electronic control, for although it is unlikely that electronics will compete with the pressure controlled valve for precise control at low cost, it is considerably more flexible in its ability to respond to stimuli from large numbers of variables, especially when microprocessors are included in the control system. It is worth remarking that Paul and Steimle (1980) attribute significant losses of COP at part-load operation to the poor performance of existing expansion valve designs, and especially to their failure to control superheat accurately. In view of this, the microprocessor controlled electric expansion valve may well have a future, despite the extra expense.

The simplest alternative to a thermostatic expansion valve is the capillary tube mentioned earlier. This is used in most domestic refrigerators and in many small freezers. Its operation relies on the fact that the mass flow through a given capillary under a given pressure is greater for a liquid than for a gas. Under normal operation, the refrigerant is fully liquid in the earlier part of the capillary, and becomes partly vaporised in the later part. If the mass flow of the system becomes too high for any reason, the condenser fails to condense it all and some gas enters the capillary, increasing impedance and restoring correct operation. If the mass flow is too small, liquid queues up at the entrance to the capillary and the amount of subcooling is increased, so the refrigerant remains in a liquid state further along the capillary tube than usual; the mass flow is increased, again restoring correct operation. This is ingenious, simple and cheap, but it is not able to cope with as wide a range of operating conditions as a thermostatic valve. It is also more difficult to design for optimum evaporator operation, although several graphical aids are available (see for example ASHRAE 1974). On the whole it is not desirable to use capillary expansion systems in heat pumps except on grounds of cheapness, though they have the inherent ability to reduce the starting load on the compressor to nearly zero, which may well have attractions in the quest for lower starting currents.

An interesting alternative is the so-called automatic expansion valve, also known as the constant pressure expansion valve. This is a spring controlled, diaphragm operated device rather like a thermostatic valve, but without the sensing bulb. Its function is to regulate the mass flow so as to maintain a steady pressure within the evaporator, thus balancing the compressor pumping capacity. When compared with the thermostatic valve, this might appear to be a better alternative for heat pump operation, because in cold weather the refrigerant flow is increased in an attempt to raise the evaporating temperature to the set point of the valve. In warm weather the flow is reduced as the maximum pressure is limited. This is precisely what is required. However, the penalty is that, unlike the thermostatic valve, the constant pressure valve tends to overflood the evaporator in cold weather, with consequent risk of a damaged compressor, and to underfeed it in warm weather leading to underutilisation of the heat transfer capacity of the evaporator at a time when a good coefficient of performance might otherwise be available. As will be discussed later in

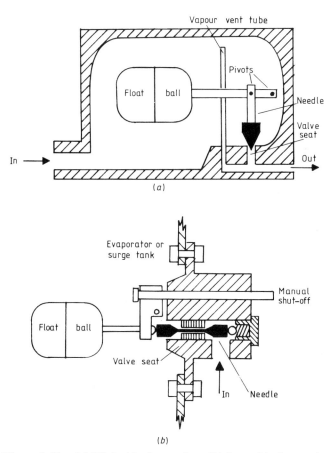

Figure 2.12 (a) High-side float valve. (b) Low-side float valve.

Chapter 4, there may be ways around this disadvantage by electronic control of evaporator capacity. There is a need for experimental study of this type of valve in comparison with thermostatic types, and it should be noted that it is possible to convert from a thermostatic to a constant pressure characteristic for experimental purposes, by the simple expedient of removing the sensing bulb from the evaporator suction line and immersing it in a constant temperature liquid bath.

Another way of achieving control of the supply of refrigerant to the evaporator is to use not a thermostatic valve but a float valve. This controls the refrigerant flow by ensuring that the level of the refrigerant in the evaporator coil is constant rather than by sensing the superheat of the refrigerant. Obviously, this achieves the same result since refrigerant is only admitted to the evaporator as it is boiled off. Float valves tend to be more expensive than thermostatic expansion valves, and also they cannot so readily be used in mobile applications where movement of the liquid might yield false indications about the refrigerant level in the system. Two examples of float valves are shown in figure 2.12. These differ only in the position of the float sensing element on the high-pressure or low-pressure sides of the orifice.

2.2 Refrigerants

The refrigerant is the working fluid of the heat pump. Heat is extracted at the cold source by evaporating the refrigerant and is rejected at the hot sink by condensing it. The properties of the refrigerant are the main factors in controlling the performance of the unit. The choice of a particular refrigerant is determined in principle by its thermodynamic properties, (for example its capacity for transferring heat) but many other factors have to be taken into account in practical situations. Dominant among these secondary considerations are cost and availability, flammability, toxicity, viscosity, surface tension and density, and the selection of refrigerant for a particular application is always a compromise between these various factors.

The ideal refrigerant will have a reasonably high pressure in the evaporator, but a relatively low pressure in the condenser, implying a low compression ratio. It will have a very high latent heat of evaporation, so that the heat transfer per unit mass of circulating refrigerant is as great as possible. In order to reduce losses in the system, the viscosity and surface tension should be small, but in this event, droplet formation during condensation will be inhibited. As this is a desirable characteristic for effective heat transfer in the condenser, obviously some compromise must be struck.

Toxicity is always a vexed problem in applications such as air conditioning, refrigeration and heat pumps. In the event of any leakage from the system, it is essential that the escaping refrigerant vapour should not injure bystanders, or contaminate food. Safety has always been considered of importance, but with the rise of emphasis on environmental protection during the 1970s, these considerations have taken on a new significance.

Indeed, only fairly recently, one of the common refrigerants R21 has been reclassified from having a maximum permissible concentration in air of 1000 parts per million (a fairly common permissible concentration) to one of 10 parts per million, after it began to be apparent that it has long term effects on the liver and also appears to be a foetal toxin. This type of consideration, together with the continuing debate as to whether or not the release of halogenated hydrocarbons affects the concentration of ozone in the upper atmosphere, (and whether or not this is of importance for the radiation balance at the earth's surface) has kept up the pressure for a continuing investigation of the properties of refrigerant gases to ensure as far as possible that those in use are 'safe'.

There is a multitude of available refrigerant materials, and they can be classified by their chemical nature. Firstly, there are inorganic compounds of which the most important are ammonia, carbon dioxide and sulphur dioxide. Secondly, there are the saturated aliphatic hydrocarbons such as methane and ethane, and thirdly, there are the unsaturated hydrocarbons such as ethylene. The fourth group is the most important class of modern refrigerants—the halogenated hydrocarbons. These form a wide range of materials, some based on the saturated hydrocarbons (for example tri-chlorofluoromethane), some on the unsaturated hydrocarbons (for example tetrafluoroethylene), and some on the cyclic organic compounds (such as octacyclofluorobutane).

Table 2.1 lists some of the common refrigerants together with their main physical properties and their conventional assigned refrigerant number. The assignation of this number is interesting in itself as it represents a systematic classification of the material and is not an arbitrary designation. For the halogenated hydrocarbons, the code consists of a three-digit number. The first digit is the number of carbon atoms less one. The second is the number of unsubstituted hydrogen atoms plus one. The third is the number of fluorine atoms. Leading zeros are not written out, so that two-digit numbers refer to methane derivatives: R12 has $(0 + 1)$ carbon atoms, $(1 - 1)$ unsubstituted hydrogen atoms and 2 fluorine atoms. Since methane has four hydrogen atoms, there must therefore be 2 chlorine atoms in R12, making it CCl_2F_2. Similarly, R22 is $CHClF_2$, and R114 is $C_2Cl_2F_4$. For cyclic hydrocarbons, a C is placed in front of the number (C318), and isomers are designated by the additional letters a, b, c depending on the extent of the deviation from the most symmetrical structure. The letter B after the figure, followed by a number, indicates that that number of bromine atoms have been substituted for chlorines (R13B1— CF_3B). Azeotropic mixtures are represented by numbers beginning with 5, for example R502. For the inorganic refrigerants, 700 is added to the molecular weight so that ammonia is R717.

Figure 2.13 shows the relationship between the various methane based possibilities, with methane (R50) at the bottom left-hand corner, carbon tetrachloride (R10) at the top and carbon tetrafluoride (R14) at the bottom right-hand corner. Figure 2.14 shows the boiling points of these compounds

Table 2.1 Physical data for some common refrigerants.

			Refrigerant		
	11	12	22	114	502
Chemical formula	CFCl$_3$	CF$_2$Cl$_2$	CHF$_2$Cl	CF$_2$Cl-CF$_2$Cl	R22 + R115
Chemical name	monofluoro-trichloro-methane	difluoro-dichloro-methane	difluoro-monochloro-methane	1,1,2,2,tetra-fluorodichloro-ethane	48.8% R22 by weight
Molecular weight	137.38	120.92	86.48	170.93	111.6
Boiling point at 1 bar (°C)	23.77	−29.80	−40.80	3.55	−45.4
Freezing point (°C)	−111	−158	−160	−94	−
Critical temperature (°C)	198.0	111.5	96.0	145.7	82.2
Critical pressure (bar)	43.74	40.09	49.36	32.66	40.75
Critical density (kg m^{-3})	554	557.6	525	582	561
Specific heat ratio C_p/C_v	1.13	1.13	1.178	1.084	—
Saturation pressure (bar)					
at 0°C	0.4	3.09	5.00	0.88	5.73
at 50°C	2.37	12.12	19.64	4.50	21.01
Heat of vaporisation (kJ kg^{-1})					
at 0°C	190.29	154.77	206.82	138.49	147.73
at 50°C	171.84	124.64	151.04	117.78	101.57
Vapour specific volume (m^3 kg^{-1})					
at 0°C	0.403	0.05538	0.04713	0.14635	0.03084
at 50°C	0.0775	0.01417	0.01167	0.03112	0.00770
Refrigerating effect (kJ m^{-3})					
Evaporation at 0°C, condensation at 50°C	364	1850.6	2949.7	605.4	2763.1
Evaporation at 10°C, condensation at 50°C	551.5	2607.1	5661.0	1652.7	3900.3

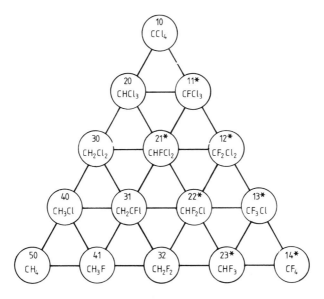

Figure 2.13 Relationship between methane based halogenated hydrocarbons. The asterisks indicate the common refrigerants.

and indicates clearly that the substitution of chlorine elevates the boiling
point and that fluorine decreases it again. The manufacturing procedure
is to start with methane, progressively chlorinating it to produce methyl
chloride, methylene chloride, chloroform and carbon tetrachloride. These
products are then fluorinated to produce the other members of the group.
A similar diagram can be constructed for the ethane derivatives etc.

The chlorine derivatives of methane without fluorine have a toxic and
narcotic effect, but this toxicity decreases with increasing fluorine content.
In addition, the chemical stability is improved as the number of fluorine
atoms increases. Higher fluorine content results in a lower temperature of
adiabatic compression, reduces the swelling action on rubber and other
elastomers and decreases the solvent power for mineral oils. Compounds
near the apex of figure 2.13 have higher molecular weight and specific
gravity; those toward the base are associated with a higher refrigerating
effect per unit of swept volume in the compressor, and with a lower
compression ratio.

One reason why the halogenated hydrocarbons are so widely used today
is their comparative safety. Of the most popular inorganics, ammonia forms
an explosive mixture with air at concentrations of about 20% by volume,
and is, in addition, toxic. A concentration of 1% can be fatal within one
hour. Sulphur dioxide is non-flammable, but is even more toxic, the same
concentration being lethal within minutes. The hydrocarbons are not toxic,
but are all flammable and prone to explosions. By contrast, the halogenated
hydrocarbons are relatively safe. There are some, such as methyl chloride,

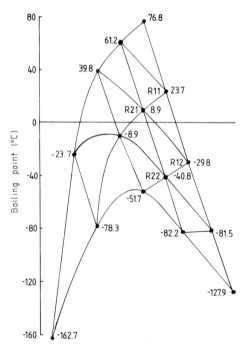

Figure 2.14 Boiling points of methane based halogenated hydrocarbons.

that are inflammable; there are some such as carbon tetrachloride and
chloroform whose toxicity approaches that of ammonia, but most are
considerably better than this and chlorodifluoromethane (R22) and tri-
chlorofluoromethane (R11) are classified as being as safe as carbon dioxide.
Safest of all are dichlorodifluoromethane (R12) and bromotrifluorome-
thane (R13B1) which are regarded as non-toxic in concentrations of up to
20%.

Another factor favouring the halogenated hydrocarbons is their chemical
stability. Ammonia is corrosive and cannot be used with copper or copper
alloys. Sulphur dioxide will dissolve in water to form sulphurous acid,
which can attack iron and steel. By contrast, the halogenated hydrocarbons
can generally be used with most common metals. Metals do tend to promote
the thermal decomposition of these compounds, but this is not usually a
problem under normal operating conditions. It will require examination,
however, as the range of application is extended to higher temperatures.

Knowledge of the thermodynamic properties of the refrigerant is important for the investigation of heat pump behaviour. Thus, it is important for
research purposes to have available accurate equations of state which will
allow computer calculations of the thermodynamic properties over a wide
range of operating conditions. An equation of state is a mathematical
relation between pressure P, specific volume v and the temperature T of
the system when it is in thermodynamic equilibrium. The form most

commonly used (at least in an underlying form) for the vapour phase of the fluorinated hydrocarbons is the Martin–Hou equation (Martin and Hou 1955), which is the basis of the commonly available tables and charts of thermodynamic properties. This equation is

$$P = \frac{RT}{v-b} + \frac{A_2 + B_2T + C_2\exp\left[-(kT/T_c)\right]}{(v-b)^2}$$
$$+ \frac{A_3 + B_3T + C_3\exp\left[-(kT/T_c)\right]}{(v-b)^3}$$
$$+ \frac{A_4 + B_4T}{(v-b)^4} + \frac{A_5 + B_5 + C_5\exp\left[-(kT/T_c)\right]}{(v-b)^5}$$
$$+ (A_6 + B_6T)e^{-\alpha v},$$

where $A_2, B_2, C_2, A_3, B_3, C_3, A_4, B_4, A_5, B_5, C_5, A_6, B_6, k, b$ and α are constant coefficients to be determined for each refrigerant.

The primary data available for determining these constants and the thermodynamic properties are:

(a) $P–v–T$ values over the range of application;
(b) heat capacity values (C_p or C_v) along one isobar or isometric for the range of temperature involved;
(c) vapour pressure values; and
(d) critical point data.

Using these and the equations of state, entropies and enthalpies, etc, can be determined.

The first of these properties is entropy. Differences in entropy can be calculated using the Maxwell relations and the $P–v–T$ data:

$$\left(\frac{\partial s}{\partial P}\right)_T = -\left(\frac{\partial v}{\partial T}\right)_p \qquad \text{and} \qquad \left(\frac{\partial s}{\partial v}\right)_T = \left(\frac{\partial P}{\partial T}\right)_v.$$

They can also be calculated from heat capacity data since

$$c_p = \left(\frac{\partial h}{\partial T}\right)_p \qquad \text{and} \qquad c_v = \left(\frac{\partial u}{\partial T}\right)_v.$$

However, $dh = Tds + vdP$, so

$$c_p = T\left(\frac{\partial s}{\partial T}\right)_p.$$

Similarly, $du = Tds - Pdv$, so

$$c_v = T\left(\frac{\partial s}{\partial T}\right)_v.$$

Thus, values of entropy changes along isobars or isometrics may be obtained from the c_p or c_v data.

During a phase change, changes in entropy may be calculated from vapour pressure data and $P–v–T$ data at the phase boundaries using the

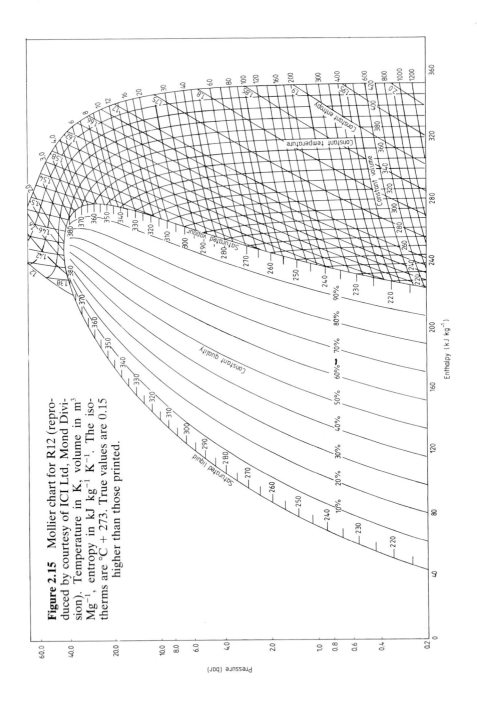

Figure 2.15 Mollier chart for R12 (reproduced by courtesy of ICI Ltd, Mond Division). Temperature in K, volume in m³ Mg⁻¹, entropy in kJ kg⁻¹ K⁻¹. The isotherms are °C + 273. True values are 0.15 higher than those printed.

46

Clapeyron equation

$$\frac{\mathrm{d}P}{\mathrm{d}T} = \frac{s_g - s_l}{v_g - v_l},$$

where g refers to the saturated vapour phase and l implies saturated liquid. The slope of the vapour pressure curve is $\mathrm{d}P/\mathrm{d}T$ and this relation is valid for first-order changes. These five equations, therefore, provide the basis for constructing an entropy table, and an example of the procedure is given in Chapter 1 of ASHRAE (1974).

Once the entropy table is constructed, internal energy and enthalpy can be calculated by using the relations

$$\mathrm{d}s = \frac{\mathrm{d}u}{T} + \frac{P}{T}\mathrm{d}v$$

and

$$\mathrm{d}s = \frac{\mathrm{d}h}{T} - \frac{v}{T}\mathrm{d}P$$

together with the expressions $c_p = (\partial h/\partial T)_p$ and $c_v = (\partial u/\partial T)_v$ and the Clapeyron equation

$$\frac{\mathrm{d}P}{\mathrm{d}T} = \frac{h_g - h_l}{T(v_g - v_l)}.$$

The principal errors in these determinations arise from the use of calorimetric specific heats, and these may be improved by using zero-pressure specific heats determined from spectroscopic data and extending the isothermal paths to $P = 0$.

A number of computer programs have been written to calculate the refrigerant properties, and one version, used in the authors' laboratory, is listed in Appendix 1. This suite of programs is designed for refrigerants R11, R12, R22, R502 and R114, and allows for the calculation of refrigerant properties in vapour or liquid phases, and in the boiling range if the enthalpy or quality of the vapour is known. The programs are based on the earlier work of Downing (1972) and Kartsounes and Erth (1971). A sample Mollier chart for refrigerant R12 is shown in figure 2.15.

2.3 Analysis of heat pump systems

As was pointed out in Chapter 1, the performance of a real heat pump is disappointingly low when compared with the ideal of the reversed Carnot cycle. Part of this can be accounted for by the fact that the practical cycle depends on latent heat transfer so that the appropriate basic process is represented by the Rankine cycle, and part arises through the appearance of irreversibilities in the thermodynamic processes.

The Rankine cycle heat engine involves condensation of the working fluid and is another idealised cycle which is unattainable in practice. It can

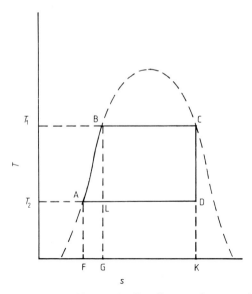

Figure 2.16 Rankine cycle T–s diagram (no superheat).

be represented on a T–s diagram as shown in figure 2.16. Initially the working fluid is at point A, with temperature T_2. It is heated to temperature T_1 along the line AB. The gain in entropy is FG $= c_p \ln (T_1/T_2)$. Next, the fluid is fully vaporised at temperature T_1 along BC yielding an entropy gain GK $= L_1/T_1$, where L_1 is the latent heat of evaporation at temperature T_1. The fluid is now expanded adiabatically along CD to temperature T_2 and condensed back to A once more. The total work done during the cycle is given by the area ABCD and the heat supplied by FABCK. The efficiency is therefore

$$\eta = \frac{\text{Area ABCD}}{\text{Area FABCK}} = \frac{\text{FABG} - \text{FALG} + \text{LBCD}}{\text{FABG} + \text{GBCK}}$$

$$= \frac{c_p (T_1 - T_2) - c_p T_2 \ln (T_1/T_2) + (L_1/T_1)(T_1 - T_2)}{c_p (T_1 - T_2) + L_1}.$$

Another possibility is that the working fluid is superheated. In this case the cycle is as shown in figure 2.17 (compare with figure 1.4) and it contains an extra section KCMR. The increase in work due to the superheating is DCMN and the extra heat supplied is KCMR. The efficiency of this part of the cycle is higher because the superheat temperature T_3 is greater. The total work done in the cycle is now ABCMN and is given by

$$W = c_p^l (T_1 - T_2) - c_p^l T_2 \ln (T_1/T_2) + L_1(T_1 - T_2)/T_1 + c_p^v(T_3 - T_1)$$
$$- c_p^v T_2 \ln (T_3/T_1),$$

where c_p^l and c_p^v represent the specific heats at constant pressure in the liquid and vapour phases. The heat supplied is

48

$$Q = \text{FABCMR} = \text{FABCK} + \text{KCMR} = L_1 + c_p^l(T_1 - T_2) + c_p^v(T_3 - T_1),$$

and the efficiency is W/Q. The evaluation of this is complicated and the alternative approach in terms of the pressure–enthalpy chart is clearly attractive as the necessary terms can be read off the horizontal axis. For heat pump applications, the cycle is reversed and the expressions for COP_h follow accordingly.

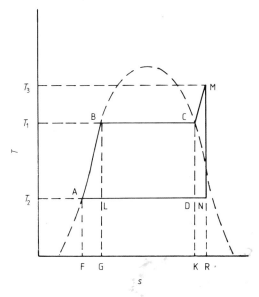

Figure 2.17 Rankine cycle $T–s$ diagram (with superheat).

The Rankine cycle expression clearly introduces the properties of the working fluid itself into the COP calculation and gives our first indication that some materials might be preferable to others. For example, in an ideal reversed Rankine cycle heat pump operating between 0 and 50 °C with R12 as the working fluid (assuming no subcooling at the condenser, no superheating at the evaporator and isentropic compression), reference to table 2.2 shows that the discharge temperature from the compressor will be 56.7 °C. The work of compression will be 24.4 kJ kg^{-1}, and the COP_h will be 5.2. If, by contrast, refrigerant R11 is used, these will become 60 °C, 31.06 kJ kg^{-1} and 5.68 respectively. The Carnot cycle heat pump between the same two temperatures would have a COP_h of 6.464 and the discharge temperature would, of course, be 50 °C.

Thus, in a reversed Rankine cycle heat pump picking up heat at 0 °C the circulation of 1 g per second of R11 through the system will require 31.06 W input to the compressor and will result in 176.44 W heat being delivered at 50 °C. With R12, a mass flow rate of 1 g per second will require a compressor power of only 24.4 W, but will deliver only 127.01 W heat at

Table 2.2 Comparison of R11 and R12. (Reference points are identified in figure 2.18.)

Property		State point				
		N†	M	C	B	L
Temperature (°C)	R11	0	59.97	50	50	0
	R12	0	56.65	50	50	0
Pressure (bar)	R11	0.4019	2.3471	2.3471	2.3471	0.4019
	R12	3.0869	12.1964	12.1964	12.1964	3.0869
Enthalpy (kJ kg⁻¹)	R11	222.90	253.96	247.70	77.52	77.52
	R12	187.53	211.95	206.45	84.94	84.94
Entropy (kJ kg⁻¹ k⁻¹)	R11	0.8260	0.8260	0.8069	0.2803	0.2938
	R12	0.6965	0.6965	0.6797	0.3037	0.3210
Specific volume of vapour (m³ kg⁻¹)	R11	0.40309	0.08066	0.07763	—	—
	R12	0.05538	0.01483	0.01417	—	—

Other factors	R11	R12
Pressure ratio	5.84	3.95
Percentage of total heat output appearing as superheat (M–C)	3.5	4.3
Quality of vapour at inlet to evaporator (L)	0.230	0.323

† In figure 2.17 N is taken as lying on the saturation line.

50°C. This makes it appear that R11 is the preferred refrigerant. Further inspection of table 2.2, however, shows the difficulties. Firstly, the compression ratio is higher with R11, and the evaporator pressure is subatmospheric. More importantly, however, the specific volumes of the two refrigerants at the inlet to the compressor are very different. R12 requires 0.05538 m^3 kg^{-1}, while R11 occupies 0.40309 m^3 kg^{-1}. This is a serious difference since in any practical system the compressor displacement is fixed and is determined by the cylinder size and the piston speed (or equivalent). Thus, a compressor with a volumetric displacement of 0.26 m^3 per minute, with R12 as the working fluid, will deliver 10 kW of heat at 50°C when the evaporator is at 0°C requiring a power input of 1.92 kW. By contrast, if R11 is the working fluid, the same compressor will deliver only 1.9 kW heat, but will require a power input of only 0.33 kW. For the same heat output, an R11 system requires a compressor with a displacement some five times greater than its R12 equivalent—even though the compressor power required is about 9% less.

This analysis, based on the Rankine cycle and the physical properties of the actual refrigerant used, is very helpful as it gives a measure of the best possible performance with that particular refrigerant. It takes no account, however, of the performance of the system components themselves, of the irreversibilities introduced through friction, the non-adiabatic behaviour of the compressor or the finite temperature difference effects at the heat exchangers. In order to account for these and to be able to assess accurately the behaviour of a particular installation, the Second Law of Thermodynamics and the concept of available energy must be introduced.

Available energy is a straightforward concept. It is that energy in a given system which is available for the production of mechanical work. Thus, a quantity of heat Q at temperature T has an available energy of $E_a = Q(T - T_0)/T$, where T_0 is the temperature of the coolest naturally available heat sink. Following this, the unavailable energy is $E_u = Q - E_a\,QT_0/T$. Any irreversible thermodynamic process will result in the degradation of the available energy, that is, less work will be available than if the irreversibility had not been present. The *irreversibility* I is the amount of the available energy that is rendered unavailable by the irreversibilities, and it is a straightforward matter to show that for an isolated system:

$$I = T_0\,(\Delta S_{\text{sink}} + \Delta S_{\text{source}}) = T_0\Delta S_{\text{total}},$$

where T_0 is the temperature of coldest naturally available heat sink, and

$$\Delta S_{\text{sink}} = Q_1/T_1, \qquad \Delta S_{\text{source}} = Q_2/T_2.$$

Here, T_1 is the temperature of the heat sink, T_2 is the temperature of the heat source, and quantities of heat Q_1 and Q_2 are transferred as appropriate. It is probably worth noting that in most heat pump applications, the *source* temperature is the same as that of the coolest naturally available heat sink. This is in contrast with the analysis of the refrigerator, where the *sink* temperature (usually the atmosphere) is equal to T_0.

In order to analyse actual heat pump cycles, we must introduce nomenclature to avoid (or at least minimise) confusion as to what is actually being referred to at any moment. In the following analysis

Q_r = heat absorbed at the evaporator (refrigerated space),
Q_s = heat rejected at the condenser (heat sink),
W = mechanical energy supplied to the compressor,
W_{id} = energy required for an ideal cycle (Carnot),
$\Delta W = W - W_{id}$,
T_r = source temperature,
T_s = sink temperature,
ΔS_s = entropy change at sink,
ΔS_r = entropy change at source,
E_{ar} = available energy at source
E_{as} = available energy at sink
E_{ur} = unavailable energy at source
E_{us} = unavailable energy at sink
E_{adr} = available energy degraded at source,
E_{ads} = available energy degraded at sink,
E_{adR} = the available energy degraded by the refrigerant in its cyclical operation.

In actual cycles, the process of heat transfer through finite temperature differences causes degradation of the available energy, an increase in the entropy of the isolated system and an augmented work input. Now,

$$W = Q_s - Q_r,$$
$$W_{id} = (T_s - T_r) Q_s / T_s = (T_s - T_r) \Delta S_s,$$
$$Q_s = T_s \Delta S_s, \qquad Q_r = T_r (- \Delta S_r).$$

Therefore,

$$\Delta W = T_s \Delta S_s - T_r (- \Delta S_r) - (T_s - T_r) \Delta S_s$$
$$= T_p (\Delta S_s + \Delta S_r) = T_r \Delta S_{total}.$$

Also,

$$Q_s = E_{as} + E_{us}, \qquad E_{us} = E_{ur} + E_{adr} + E_{adR},$$
$$E_{ur} = T_r (- \Delta S_r), \qquad E_{ads} = E_{as} - W_{id}$$

and so,

$$\Delta W = W - W_{id},$$
$$= Q_s - Q_r - (T_s - T_r) \Delta S_s,$$
$$= E_{as} + T_r (- \Delta S_r) + E_{adR} + E_{adr} - T_r (- \Delta S_r) - T_r (T_s - T_r) \Delta S_s,$$
$$= E_{ads} + E_{adr} + E_{adR}$$
$$= E_{ad},$$

where E_{ad} is the total available energy degraded in the cycle.

If any analysis is to be performed on an actual heat pump system, then it is apparent that the state point properties of the refrigerant must be known at key points around the cycle. As a minimum requirement, they must be known at the inlet of every major component and hence, by implication, at the outlet of the preceding component. Ideally, they should be available at both the inlet and outlet of every component so that the influence of the connecting pipes can also be assessed.

To illustrate the analysis let us consider the following simplified set of results from a real installation. The simplification was made by assuming (a) that the pipe runs from the compressor to the condenser and from the condenser to the expansion valve, forming part of the condenser, (b) that the refrigerant liquid spreader, included at the inlet to the evaporator to ensure an even distribution of refrigerant among the different evaporator circuits, is part of the expansion valve, and (c) that the suction pipe between the evaporator and the compressor is part of the evaporator. The experimental data used were the measured pressures and temperatures at the appropriate points. The actual points are indicated in figure 2.18, and the state point properties calculated from our suite of programs are given in table 2.3. Point 7 is included to show the influence of heat exchange between secondary system components and the outside air.

Taking the basic four-point cycle 1468, we can readily write down the expressions for the overall system performance:

$$Q_r = h_8 - h_6 = 98.5 \text{ kJ kg}^{-1},$$
$$Q_s = h_1 - h_4 = 131.1 \text{ kJ kg}^{-1},$$
$$W = Q_s - Q_r = 32.6 \text{ kJ kg}^{-1},$$
$$\text{COP}_h = 131.1/32.6 = 4.02,$$
$$\text{COP}_{id} = T_{sink}/(T_{sink} - T_{source}) = 325/(325 - 280) = 7.222,$$
$$W_{id} = 131.1/\text{COP}_{id} = 18.152308 \text{ kJ kg}^{-1},$$
$$\Delta W = W - W_{id} = 14.447692 \text{ kJ kg}^{-1}.$$

Thus, we see that this particular heat pump was rejecting 131.1 kJ for each kg of refrigerant circulated around the system, and required 32.6 kJ of mechanical energy to achieve this. The ideal work that would have been necessary to achieve the same heat rejection was 18.152308 kJ, showing that an extra energy input of 14.447692 kJ had been required. This value *should* be the same as $T_r \Delta S_{total}$ since T_r is the temperature of the lowest naturally available heat source. To check this we must determine the change in entropy of the source and sink and then calculate ΔS_{total}.

At the source, 98.5 kJ kg^{-1} is removed, so that the entropy is changed by an amount $\Delta S_r = -98.5/280 = -0.3517857 \text{ kJ kg}^{-1} \text{ K}^{-1}$. At the sink the delivery of heat by the heat pump yields an entropy change $\Delta S_s = 131.1/325 = 0.4033846 \text{ kJ kg}^{-1} \text{ K}^{-1}$, whence $\Delta S_{total} = 0.0515989 \text{ kJ kg}^{-1}$ K^{-1}, and $\Delta W = T_r \Delta S_{total} = 280 \times 0.0515989 = 14.447692 \text{ kJ kg}^{-1} \text{ K}^{-1}$. Thankfully, we note that this agrees with our earlier value determined

Table 2.3 State point data for sample heat pump run. $T_r = T_{source}$ (air) $= 280$ K, $T_s = T_{sink}$ (water) $= 325$ K, refrigerant is R12.

State point	Pressure (bar)	Temperature (°C)	Enthalpy (kJ kg^{-1})	Entropy (kJ kg$^{-1}K^{-1}$)
1	15.128	77.0	224.1	0.7205
4	14.452	57.5	93.0	0.3277
6	3.162	0.7	93.0	0.3498
7	2.928	3.0	189.8	0.7081
8	2.928	5.7	191.5	0.7143

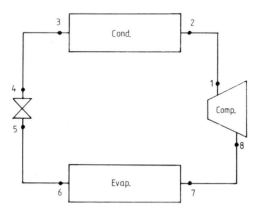

Figure 2.18 Heat pump cycle diagram numbered to identify measured state points.

from the difference between the theoretical work and the actual work.

It is worth noting, perhaps, that this is not the COP$_h$ that is actually measured on this heat pump in terms of the input electrical power. The experimental unit had a heat output of 10 kW. This implies from the thermodynamic data above, that the power input should have been 2.49 kW. Unfortunately, the electrical motor driving the compressor had an efficiency of only 80%, and in addition, the fan power needed to move the air over the evaporator was 150 W. Combining these factors produced an electrical power input of 3.26 kW which makes the real COP$_h$ = 3.07 in terms of heat out per unit energy in to the unit. Sadly, this is a more realistic practical figure for heat pumps of this capacity.

Turning back to the performance of individual components of the heat pump system, these can be evaluated in terms of the entropy change in the refrigerant and the entropy change in the surroundings because of the behaviour of each individual component. Basically, the irreversibility in the refrigerant is determined by the product of the entropy change in the refrigerant between the inlet and outlet to the component and the temperature of the heat source ($T_r \Delta S_{ref}$). The irreversibility due to the entropy

54

change in the surroundings is given by the product of the heat transferred to the surroundings and the ratio of the source temperature to the temperature of the surroundings. Both expressions have to be determined for heat exhangers, pipe lengths etc, where heat transfers are occurring, while for the expansion valve, and ideally the compressor, only the irreversibility in the refrigerant is appropriate since no heat transfer to the surroundings occurs. In our simple four point analysis therefore, the irreversibilities are given by:

$$I_{comp} = T_r\,(s_1 - s_8) = 280\,(0.7205 - 0.7143) = 1.736\ \text{kJ kg}^{-1},$$
$$I_{exp} = T_r\,(s_6 - s_4) = 280\,(0.3498 - 0.3277) = 6.188\ \text{kJ kg}^{-1},$$
$$\begin{aligned} I_{cond} &= T_r\,[s_4 - s_1 - (h_4 - h_1)/T_{sink}] = 280\,[0.3277 - 0.7205 \\ &\quad + (224.1 - 93.0)/325] = 2.963692\ \text{kJ kg}^{-1}, \end{aligned}$$
$$\begin{aligned} I_{evap} &= T_r\,[s_8 - s_6 - (h_8 - h_6)/T_r] = 280\,[0.7143 - 0.3498 \\ &\quad - (191.5 - 93.0)/280] = 3.560\ \text{kJ kg}^{-1}, \end{aligned}$$
$$I_{total} = I_{comp} + I_{exp} + I_{cond} + I_{evap} = 14.447692\ \text{kJ kg}^{-1}.$$

Once again, the total thermodynamic loss equals the extra work that has to be done in comparison to the ideal cycle, but now we have some indication of the relative importance of the individual components. This can be made into a finer sieve by providing a finer mesh of data points. For example, if point 7 is introduced at the outlet from the evaporator, then we can directly assess the influence of the suction pipe as

$$I_{pipe} = T_r\,(s_8 - s_7 - (h_8 - h_7)/T_r) = 0.036\ \text{kJ kg}^{-1}.$$

In this case, the suction pipe is having only a small effect, but this is not always so. It is important to notice that when this finer mesh is used, the actual loss in the evaporator is more accurately assessed as $3.524\ \text{kJ kg}^{-1}$, so that the calculated irreversibilities are redistributed around the system as more components are introduced in the analysis.

We can write these irreversibility losses a little differently again if we wish to express them in terms of the available energy degraded at the evaporator, the condenser, and in the refrigerant. Thus, at the evaporator

$$E_{ur} = T_r\,\Delta S_{evap} = 102.06\ \text{kJ kg}^{-1}\ \text{(four state point model)},$$
$$E_{ar} = Q_r - E_{ur} = 98.5 - 102.06 = -3.56\ \text{kJ kg}^{-1},$$
$$E_{adr} = -E_{ar} = 3.56\ \text{kJ kg}^{-1} = I_{evap}.$$

At the condenser,

$$E_{us} = T_r\,\Delta s_{cond} = 109.984\ \text{kJ kg}^{-1},$$
$$E_{as} = Q_s - E_{us} = 131.1 - 109.984 = 21.116\ \text{kJ kg}^{-1},$$
$$E_{ads} = E_{as} - W_{id} = 2.963692\ \text{kJ kg}^{-1} = I_{cond}.$$

In the refrigerant,

$$\begin{aligned} E_{adR} &= T_r\,(\Delta s_{comp} + \Delta s_{exp}) = T_r\,(s_1 - s_8 + s_6 - s_4) \\ &= 7.924\ \text{kJ kg}^{-1} = I_{comp} + I_{exp}. \end{aligned}$$

Once again, the total irreversibility is equal to the sum of the degraded available energies and to the extra work, as expected.

This type of analysis can be extended to give a very clear picture of the behaviour of each component in a system, and of course, overall consistency of the data can be obtained by comparison with the known power input to the system, and the measured heat output.

Thermodynamic analysis really demonstrates its usefulness when showing the relative performance of various components in a particular machine and the sensitivity of their performance to changes in operating conditions or design parameters. For example, in the case of table 2.3, it is apparent that the largest single irreversibility is taking place in the expansion process, which appears to be accounting for twice the loss of the condenser or evaporator. Now, this expansion is actually taking place in two separate components, the expansion valve itself and the liquid spreaders at the evaporator inlet. These are present in all multi-circuit evaporators to ensure an even distribution of refrigerant supply to each circuit. Thus, part of the loss is present for practical reasons other than those directly associated with the expansion process. However, the irreversibility is present nonetheless, and is a necessary adjunct of the partial flash of vapour during expansion. Subcooling of the liquid at the condenser, already referred to in Chapter 1, reduces this phenomenon but is only achieved at the penalty of increasing the condensing temperature, and hence the discharge temperature, compressor work, etc. Whether this represents a nett advantage or disadvantage in a given case can be readily assessed using the analysis above.

One example of the use of this analysis in assessing the sensitivity of the system to component variations can be seen if the compressor efficiency is changed. (This change affects only one point in the cycle (1) and so is ideal for illustrative purposes). Assume that the compressor is less efficient, so that more work of compression is required to achieve the same discharge pressure and condensing temperature. Let us assume that point (1) is now specified by $T = 82°C$, $P = 15.128$ bar, $h = 228.19126$ kJ kg^{-1} and $s = 0.73206$ kJ kg^{-1}. The analysis can now be repeated exactly as before, the results being:

$$
\begin{aligned}
Q_r &= 98.5 \text{ kJ kg}^{-1} \\
Q_s &= 135.19126 \text{ kJ kg}^{-1} \\
W &= 36.69126 \text{ kJ kg}^{-1} \\
\Delta W &= 18.538952 \text{ kJ kg}^{-1} \\
\text{COP}_h &= 3.684563 \\
\Delta S_{total} &= 0.0662105 \text{ kJ kg}^{-1} \text{K}^{-1} \\
I_{comp} &= 4.9728 \text{ kJ kg}^{-1} \\
I_{cond} &= 3.2516702 \text{ kJ kg}^{-1}
\end{aligned}
$$

and so on for the other quantities calculated earlier. If the fan power (150 W) and electrical motor efficiency (80%) are included, this leads to a practical COP$_h$ of 2.82 for a heat output of 10 kW, compared with the previous figure of 3.07.

In a similar manner, the sensitivity of the system performance to changes in other system components can be assessed, and the consequences of design changes predicted.

2.4 Effects of lubricating oil

Lubricating oil is an essential feature of every existing refrigeration or heat pump system. The oil is needed to keep the compressor running, and also helps to seal the cylinders during the compression stroke. Unfortunately, some lubricant becomes entrained with the refrigerant and is circulated around the system. With a refrigerant such as R22, which has a very low solubility in the oil, this causes a difficulty in that the oil has to be separated from the refrigerant and returned to the oil sump of the compressor—not a difficult task. With other refrigerants such as R12, however, the miscibility is so high that this separation is impossible and the oil is swept around the system. Care must be taken, therefore, to ensure that at all points in the circuit where the refrigerant is fully vaporised the refrigerant vapour speeds are sufficiently high to entrain the non-volatile liquid lubricant. Thus, some pressure drops are forced on the system to ensure its operation. The danger in not designing these pressure losses into the system lies in the fact that oil will be trapped in the section concerned. Thus, if one circuit of the evaporator does not clear the oil adequately, it will become oil-logged and will cease to function as an evaporator. It is tempting to suggest that an oil separator at the compressor discharge should be included as a standard feature of heat pump design. Unfortunately, such an oil separator is really a *liquid* separator, and with a refrigerant such as R12, at the discharge temperatures encountered in heat pump applications, the liquid will typically comprise about 50% refrigerant (and can be frequently as high as 75%). Thus, two to three times as much refrigerant will be returned as oil. In an oil-flooded compressor design such as the rotary sliding vane compressor, with oil concentrations of about 10% circulating with the refrigerant, this would mean some 25–30% of the refrigerant being returned directly to the compressor inlet. This would entail a significant loss in performance.

The presence of the oil has other implications, and the principal effects can be summarised as follows (Hughes *et al* 1980):

(a) It changes the working fluid from a pure refrigerant with well defined properties to a poorly understood mixture with properties that depend on the oil concentration.

(b) It changes the refrigerant-side heat transfer coefficient in the evaporator and condenser, leading to either improvement or impairment of heat transfer, depending on oil concentration.

(c) The boiling point of the refrigerant oil mixture is elevated above that of the pure refrigerant for a particular pressure, according to Raoult's Law.

(d) The latent heat carrying capacity of the mixture is reduced because the oil holds a proportion of the refrigerant in the liquid phase, preventing its evaporation.

This last point is the most significant in relation to heat pumps. In refrigeration or air-conditioning plant, where the production of cooling is the prime objective, the reduction in evaporator capacity caused by the presence of oil can be compensated for by merely increasing the plant size. However, COP is the most important performance parameter of a heat pump, and the result of (d) is a reduction in the COP because the evaporator capacity is reduced without a corresponding reduction in compressor power. This is an effect that cannot be remedied by an increase in plant size.

It is fairly straightforward to show the general type of phenomenon that will be encountered when oil is circulating with the refrigerant, though any detailed understanding is hampered by the poor state of our knowledge of the properties of refrigerant–oil mixtures. Fortunately, Bambach (1955) does give data for R12-paraffinic oil mixtures, and we can extend some of the earlier work on capacity effects to suit our purposes. Interestingly, the summarised analysis that follows also illustrates how the presence of oil can explain the phenomenon whereby the COP of a given heat pump can be optimised by varying the evaporator superheat setting.

In 1972, Cooper and Mount showed that evaporator capacity is affected by the presence of oil according to the equation

$$\frac{Q_m}{Q_R} = \frac{\rho_{Sm}\,(h_{2_m} - h_{1_m})}{\rho_{SR}\,(h_{2_R} - h_{1_R})},$$

where Q = heat transferred, ρ = fluid bulk density, h = fluid specific enthalpy, m refers to oil–refrigerant mixture, R to pure refrigerant, S to compressor suction, 1 to evaporator inlet and 2 to evaporator outlet.

The denominator is determined from the published refrigerant data, but the numerator requires reference to Bambach's work (for R12). Figure 2.19 shows curves of evaporator capacity correction factor (Q_m/Q_R) for various oil fractions x and suction superheat temperatures. It can be seen that evaporator capacity is affected significantly by both oil circulation and suction superheat. Also, contrary to expectation, capacity is greatly reduced at low superheat levels when a small amount of oil is present.

In order to estimate the effect on COP, it is necessary to make an assumption about the compressor power. Two reasonable assumptions are possible: (1) compressor power is constant and unaffected by oil circulation and (2) compressor power varies proportionally with changes in suction density due to the oil circulation. Neither of these assumptions is likely to be correct, but they can be taken as representing 'best' and 'worst' cases. Both are used here for comparison.

If we can neglect heat losses, then heat pump COP$_h$ can be written as

$$\text{COP}_h = \frac{Q_E + W}{W},$$

and we can therefore write this for the oil–refrigerant mixture, in terms of the pure refrigerant COP$_h$ as

$$\frac{\text{COP}_{hm}}{\text{COP}_{hR}} = \frac{f_p f_h (\text{COP}_{hR} - 1) + 1}{\text{COP}_{hR}} \quad \text{for constant compressor power}$$

and

$$\frac{\text{COP}_{hm}}{\text{COP}_{hR}} = \frac{f_h (\text{COP}_{hR} - 1) + 1}{\text{COP}_{hR}} \quad \text{for density-proportional compressor power,}$$

where $f_p = \rho_{Sm}/\rho_{SR}$ and $f_h = (h_{2_m} - h_{1_m})/(h_{2_R} - h_{1_R})$.

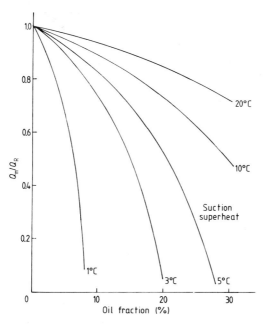

Figure 2.19 Change in evaporator capacity with oil concentration. Refrigerant, R12; evaporating temperature, 0 °C; condensing temperature, 55 °C; subcooling, 5 °C.

These relations are plotted in figure 2.20, and they can be seen to be broadly similar to those on evaporator capacity, but the effect of an optimum evaporator superheat setting is very clearly illustrated. This effect agrees with experimental observations in the authors' laboratory, and it clearly represents an important phenomenon—even at low oil concentrations.

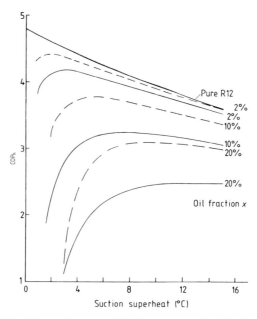

Figure 2.20 Change in COP_h with oil concentration. Refrigerant, R12; suction temperature, 5°C; condensing temperature, 55°C; subcooling, 5°C. Full curves, compressor power proportional to suction density; broken curves, constant compressor power.

2.5 Alternative thermodynamic cycles

The discussion so far would suggest that the reversed Rankine cycle is the only thermodynamic cycle currently being evaluated for heat pump applications. This is far from the case, and at least one other—the Lorenz cycle—is being actively pursued (Rojey *et al* 1980). The literature also refers to Stirling cycle and Brayton cycle heat pumps, but in both these cases closer reading indicates that it is the *drive* that is operating on the alternative cycle and not the heat pump itself. This is illustrated in Chapter 3.

The alternative thermodynamic cycle that is showing most promise for the heat pump itself is the Lorenz cycle which is based on using a mixture of refrigerants as the working fluid rather than a single pure refrigerant. This non-azeotropic mixture does not have a well defined boiling point, but rather a boiling range, and so it is possible in principle to design an evaporator in which the boiling point of the refrigerant varies in step with the temperature evolution of the external heat source.

The Lorenz cycle is shown in figure 2.21 and it is possible to determine the maximum coefficient of performance which can be obtained, provided we assume that the specific heats of the external heat exchange fluids are

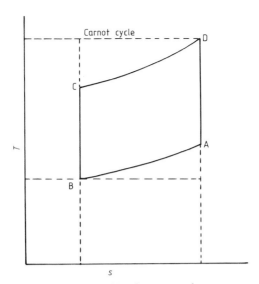

Figure 2.21 Lorenz cycle.

constant over the temperature ranges of interest. If this is so, then

$$\text{COP}_{hL} = \frac{\overline{T_1}}{\overline{T_1} - \overline{T_0}}$$

where $\overline{T_0}$ and $\overline{T_1}$ are the log mean temperatures at the evaporator and condenser respectively. That is,

$$\overline{T_0} = \frac{T_A - T_B}{\log (T_A/T_B)},$$

and

$$\overline{T_1} = \frac{T_D - T_C}{\log (T_D/T_C)}.$$

For a Carnot cycle, obviously, $\text{COP}_c = T_D/(T_D - T_B)$, and figure 2.22 shows the variation of $\text{COP}_c/\text{COP}_{hL}$ for a range of values of T_A, T_B and T_D, T_C. Comparison of this curve with the Rankine cycle values in § 2.3 would suggest that significant improvements in performance are possible, at least in principle. For example, with $T_A = 0\,^\circ\text{C}$, $T_B = -5\,^\circ\text{C}$, $T_C = 40\,^\circ\text{C}$ and $T_D = 50\,^\circ\text{C}$ the COP_h becomes 6.7 compared with the earlier value of 5.2 or 5.68 depending on the refrigerant. Thus, with the sort of temperatures that are typically found in with air–water heat pumps, and allowing for part of the temperature variation that is always present (but not any losses on the refrigerant side) a very high COP_h is possible in principle. The performance appears to be better than that from a Carnot cycle machine, but this is illusory. The Carnot cycle performance is calculated on a 'worst case' basis taking the highest condenser temperature and the lowest evaporator temperature as the reference. In the Lorenz cycle, most of the heat transfer is taking place at intermediate temperatures. However, since for

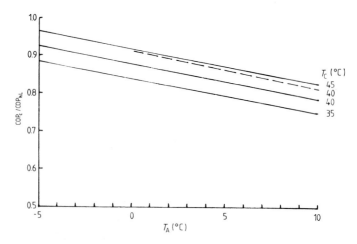

Figure 2.22 Ratio of Lorenz cycle COP to Carnot cycle COP. Full lines, $T_D = 50°C$, $T_B = -5°C$; broken line, $T_D = 50°C$, $T_B = 0°C$.

a given set of 'final' heat source and sink conditions, the ideal Rankine cycle is worse than the Carnot cycle, and the ideal Lorenz cycle is better, there is a suggestion that significant savings are possibly to be made.

Practically, an approach to Lorenz cycle operation can be achieved by using counterflow heat exchangers so that the refrigerant fluid is hottest when it is interacting with the hottest part of the external heat exchange medium.

Work on Lorenz cycle machines is being carried out in several places and in particular at the Institut Français du Petrole. Indications are that the COP might be improved by up to 50% compared to present designs, but that this will only be achieved with a doubling or more of the evaporator size. Once again, development is somewhat hampered by our limited knowledge of the properties of refrigerants and refrigerant mixtures, but the prospects for this approach seem good.

3

DRIVES FOR COMPRESSION HEAT PUMPS

3.1 Motive power

Almost all existing heat pumps, whether commercial machines or research units, use an electric motor to drive the compressor. The practical convenience of electric drives, especially for small domestic machines, is obvious: easy starting and stopping, low noise, minimal maintenance and so on. For research work there are similar advantages in electric power, coupled with the major factor that the measurement of power input is extremely easy and accurate. However, there are some significant objections to the use of electric motors, and some significant advantages to be gained by considering the use of alternative sources of motive power.

The main technical problems of electric motors are that they are often fixed-speed machines, which makes it difficult to vary the heat output of an electric heat pump by more than a small percentage, and also that in starting an electric motor of reasonable size there is a short period of heavy current demand which can create problems for other electrical equipment in the vicinity. There is also a feeling among electricity supply authorities that the load factor of heat pump heating systems is poor. It is possible to overcome these problems to some extent, and some methods for doing so will be discussed in the present chapter.

There is an important energetic objection to electrically driven heat pumps. In present-day power generation it is usual to achieve an overall efficiency of only about 30–35%, the rest of the energy being thrown away into the sea or the atmosphere. If one operates a heat pump at a COP of 3.0, with electricity generated at an efficiency of 33%, the overall efficiency with which the fuel is used, often termed the primary energy ratio, is the apparently uninteresting figure of unity. On the other hand, if one uses a fuel-fired engine to drive the compressor, one can recover a useful proportion of the waste heat of the engine, so that even if the efficiency of the engine is rather lower than that of the power station, the primary energy ratio may well be greater than unity. Of course, one can envisage radically more efficient power generation systems than are in use at present, perhaps restoring electric systems to a position of favour with conservationists, but such a change seems unlikely in the immediate future.

The case against the electric heat pump, and in favour of directly fired fuel-burning machines, becomes less convincing when one remembers that one is likely to be using a more expensive fuel to drive the engine than to fuel the power station with which it is competing. Furthermore, the engine may well use oil or natural gas, of which there is likely to be a shortage in the future, while the power station may well use coal or nuclear fuel, of which there is less likely to be a shortage, or even one of the natural power sources based on solar energy, which in the absence of an astronomical cataclysm will go on indefinitely.

It is probable that neither electric nor fuel-burning drives have all the advantages for every circumstance in which compression heat pumps are likely to be used. It is the purpose of this chapter to examine the characteristics of various alternative motive power systems, in order to clarify the choice which is available to the designer.

3.2 Electric motors

For hermetic and semihermetic compressors (see Chapter 2) the electric motor is the only practicable motive power at present. Furthermore, only certain types of electric motors are suitable. The squirrel-cage induction motor is the favourite, because it is brushless and thus avoids the problem of brush maintenance. It also avoids the possibility of the refrigerant or the oil being decomposed by sparks. For open compressors, the possibilities are more extensive, since one can gain access to brushgear for replacement when necessary, but the maintenance-free advantages of the squirrel-cage induction motor have led to their almost universal adoption. Following our policy of the previous chapters in paying greatest attention to the most immediately practicable technologies, we shall confine ourselves to such motors for most of this chapter, but the reader should bear in mind the potentialities of alternatives.

The induction motor is effectively a transformer in which current is induced in the windings of the rotor by induction from the static coils or stator. The resulting current interacts with the magnetic field of the stator to produce a torque. This principle is common to all induction motors, but there are sufficient differences between three- and single-phase machines to warrant considering them separately.

3.2.1 *Polyphase induction motors*
The torque–speed characteristics of a polyphase induction motor are shown in figure 3.1. There are three regions, two of which, the braking and generating regions, are of little importance to heat pump technology and will not be considered further. The third region, the motoring region, is the essential one. It corresponds to rotor speeds from zero to a limiting speed which is equal to the rotational speed ω_m of the AC supply mains if the stator winding has only two magnetic poles, and to a fraction $1/n$ of the supply speed if the stator has $2n$ poles. The limiting speed is therefore

ω_m/n, which in the case of a two-pole motor is 3000 rpm for a 50 Hz supply, and 3600 rpm for a 60 Hz supply; for a four-pole motor ω_m/n is half these figures, and so on. Note that the torque of the motor is zero at ω_m/n so this represents the limiting speed which can never actually be reached in practice (even when running with no load) because of friction and other losses, and certainly cannot be reached when doing useful work. There exists therefore a slip between the rotation speed ω_r of the rotor, and the limiting speed ω_m/n, measured by the fractional slip s where, for a two-pole motor,

$$s = \frac{\omega_m - \omega_r}{\omega_m}.$$

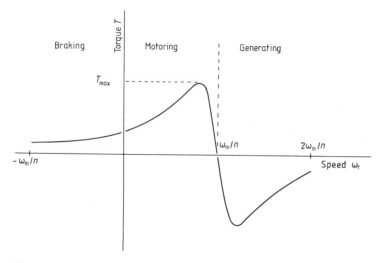

Figure 3.1 Torque–speed characteristics of a polyphase induction motor with $2n$ poles for a mains rotation speed of ω_m. At rotor speeds ω_r between zero and the limiting speed ω_m/n the machine acts as a motor. If driven faster than the limiting speed, the machine acts as an induction generator. When the mains supply is reversed, it acts as a brake.

This slip is entirely necessary to produce a torque and, indeed, the torque increases as the slip increases until the maximum is reached at T_{max} (figure 3.1). Thus as the load increases and causes the motor to slow down, the torque increases and compensates for the increased load, and the motor draws an increased current from the supply. This is a highly useful characteristic, because it allows the motor to regulate itself, taking only enough power from the supply to do the work required of it. Further increase in slip beyond T_{max} causes the torque to reduce, and this leads to a mechanically unstable condition in which the motor is unable to sustain the load and will come to a halt unless some remedial action is taken. Obviously, therefore, it is necessary to operate the motor–load combination at a speed between that corresponding to T_{max} and the limiting speed.

It is desirable to minimise the fractional slip s because the efficiency of a perfect loss-free induction motor cannot exceed $(1 - s)$. In a real motor in which losses occur due to resistance of the coils, magnetic hysteresis, air drag and so on, the actual efficiency will be less than this, but the principle still applies. This would appear to indicate that the use of a grossly oversized motor, with a correspondingly small value of slip, would make for the best efficiency. However, at zero slip the torque is also zero, so the motor does no useful work, while the supply is still called upon to make up the losses, so the efficiency will be zero at zero slip. Similarly we can expect zero efficiency at zero speed. This leads to an efficiency–speed curve as shown in figure 3.2(*a*), and the torque–speed curve for the same range as shown in figure 3.2(*b*). For clarity and simplicity we have omitted the generating and braking regions in these diagrams.

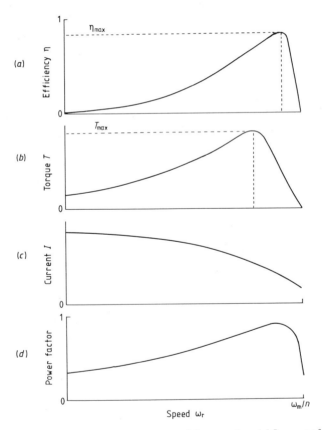

Figure 3.2   (*a*) Efficiency, (*b*) torque, (*c*) current and (*d*) power factor of a polyphase induction motor as functions of rotor speed.

Another factor of importance in motor characteristics, of especial interest to the electricity supply undertaking, is the current–speed curve shown in figure 3.2(c). In most applications it is the starting current which is of greatest importance, because at small values of ω_r the current can be considerably higher than the normal running current. It may be necessary to restrict the starting current by methods such as star–delta switching, in which the windings are connected across the phase-to-neutral voltage during starting, and are then reconnected across the higher interphase voltage when the motor has reached a suitable speed.

The power factor, shown in figure 3.2(d), is the final factor of importance, mainly because electricity supply authorities often make a higher charge for loads with poor power factor. Although the domestic consumer has traditionally been spared this imposition (mainly because domestic loads are usually nearly purely resistive), a proliferation of large heat pump motors in residential areas might cause the undertakings to rethink their policy. Furthermore, operation of a motor under conditions of poor power factor, which in practice means under low load, tends to reduce motor efficiency because the windings are carrying a substantial current, leading to substantial resistive losses, although the useful work output is low. The customary method of power factor correction is to connect a capacitor into the circuit, so as to produce a leading current vector partially cancelling out the lag of the inductance. This alleviates the problem for the electricity supply authority, but does not necessarily improve the efficiency of the motor under low load. An alternative approach to the problem has been developed by electronic switching of the power to the motor in such a way as to correct the phase angle to the desired extent. According to Pascente (1979) this can reduce part-load power consumption by as much as 40–50% but it is, of course, much less effective at full load because the power

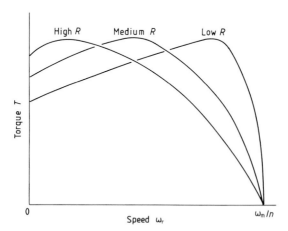

Figure 3.3 Effect of rotor resistance on torque–speed characteristic of a polyphase induction motor.

factor is usually much nearer unity under such conditions and there is accordingly much less 'useless' current.

The speed at which maximum torque is developed is influenced by the rotor resistance R, as shown in figure 3.3. Low rotor resistance, which is clearly desirable for reducing losses and maximising efficiency, also reduces the starting torque. Fortunately this is unlikely to be a problem in heat pump compressors unless an attempt is made to start the machine with a very high pressure difference across the compressor, or when the mains voltage is low ('brown-out' conditions).

From figure 3.3 it can be seen that the starting torque of a polyphase induction motor is always much greater than zero. This is simply because it is easy to create a rotating magnetic field when a set of coils is connected to a polyphase supply, and it makes such motors inherently selfstarting—a very useful attribute.

3.2.2 Single-phase induction motors

In many parts of the world it is customary to supply single-phase AC power to houses, and even though three-phase power may be available immediately outside the garden gate, it may be almost impossible, or at least very expensive, to persuade the supply authority to connect the two extra wires needed to bring the three-phase power indoors. Consequently the designer of domestic heat pumps is forced to use single-phase motors.

With only a single phase available, it is not straightforward to create a rotating magnetic field. Mathematical analysis reveals that with a single stator winding connected to a single-phase supply two torques are generated, as shown in figure 3.4, rotating in opposite directions. At zero rotor speed they are equal and cancel each other out; thus there is zero starting torque. If the rotor is given a starting speed by some means, a net torque arises, and the motor behaves thereafter rather like the three-phase versions discussed earlier. An ideally symmetrical motor should be able to rotate in either direction equally well, depending only on the initial direction of rotation.

The generation of an initial starting torque calls for a certain amount of ingenuity. The most common arrangement is to provide an auxiliary winding, wound in an orientation approximately 90° out of phase with the main winding, and to supply it with a current which is out of phase with the current in the main winding, ideally by 90°, producing what is effectively a two-phase motor. One common implementation of this arrangement is to feed one winding directly from the supply, and to feed the other via a capacitor, as in figure 3.5, which is known as the capacitor motor. The vector diagram shows that if a suitable value of capacitance is chosen, the vectors in the two phases will be at right angles, producing an ideal two-phase operation. This will be exactly true only at one particular load, and there will be some unbalance between the phases under other conditions, but not to a serious extent except during starting, when the non-ideal phase relationship leads to a rather poor starting torque. It is possible to correct for this by connecting another much larger capacitor in parallel with the existing one during starting, and then automatically switching it out when the rotor speed has reached a suitable value (see figure 3.6). The starting period is usually only a second or two and it is unnecessary for the starting capacitor to be suitable for continuous duty; it is usual to make use of an electrolytic type because of the lower cost of obtaining large capacitance with electrolytics, but it must, of course, be suitable for AC. Likewise it is unnecessary for the capacitor-fed winding to be able to withstand the large current fed via the starting capacitor for long periods, and it is usually wound with lighter gauge wire than the main winding. Both these economies are desirable, but they put a premium on reliable operation of the automatic switch, failure of which will almost certainly damage either the starting capacitor or the winding. It may also be necessary to restrict the number of starts in a given time, perhaps to about one per minute. This is unlikely to be a difficulty in normal operation, but under some fault conditions there may be a great deal of stopping and starting, and it may be necessary to incorporate an inhibiting device to prevent damage under such conditions.

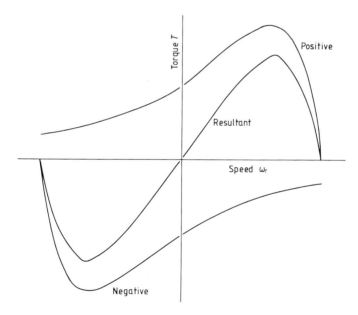

Figure 3.4   Torque–speed curve for a single-phase induction motor.

An alternative method of generating a starting torque, usually found in small motors up to about 1 kW, is to have a resistive start winding in which the inductive lead of the current vector relative to the voltage vector is less than in the main winding (figure 3.7). The start winding, being deliberately resistive, gets hot and must be switched out after a short period in a manner similar to the capacitor-start motor. Unlike it, however, it is not possible to arrange a continuous duty for the start winding. Consequently the space occupied by it is effectively useless during normal running, and this leads to a lower efficiency than for capacitor motors.

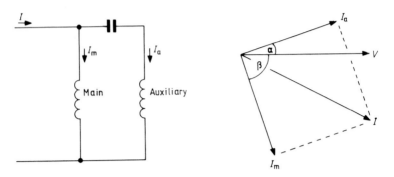

Figure 3.5 Capacitor motor. Ideally $\alpha + \beta = 90°$.

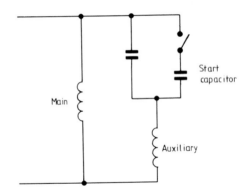

Figure 3.6 Capacitor-start, capacitor-run motor.

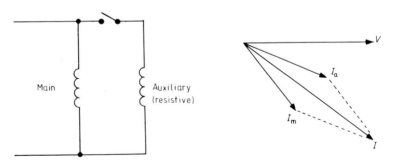

Figure 3.7 Split-phase motor with resistive start winding.

3.3 Motor starting

During the starting period, the motor is less efficient because of the lower relative speed of the stator and rotor, the induced back-emf is less (actually zero at the beginning), and electrical energy has to be consumed in speeding up the rotor against its mechanical inertia. All this inevitably means that the current is higher during the starting period. This is true of both single- and three-phase motors, but the situation is exacerbated in single-phase motors because of the need to energise a secondary winding, and the greater the starting torque required, the greater the problem becomes. A typical starting current is about five times the normal running current for a single-phase motor, and the effect of this transient on the supply voltage at the consumer's premises can be quite serious, especially for large motors. Any steps which might be taken to lessen the effect would not only improve the acceptability of heat pumps in domestic use, but also allow the use of larger machines.

There are several ways of starting a single-phase induction motor under reduced current. All of them involve reduced starting torque, and a compromise is inevitable, but the severity of reduction of torque for a given reduction of current is different for different methods. We will consider for purposes of illustration a capacitor-start, capacitor-run motor, because for most applications in heat pump compressors it is the most appropriate.

(a) The motor can be supplied via an autotransformer, and started under reduced voltage. The starting torque and starting current are both proportional to the square of the voltage tapping. Consequently if it is required to reduce the starting current to half of its direct on-line value, a tapping at $1/\sqrt{2}$ of the line voltage will be needed, and the starting torque will be half of its direct on-line value. This is a fairly good method, applicable to most motors, but is expensive to implement because of the high cost of the transformer.

(b) A series resistor may be introduced into the supply to both windings. This will reduce the voltage at the motor and thus reduce the starting current, and since the voltage reduction is directly related to the current, the effect tails off as the motor builds up speed. Nevertheless, it is desirable to switch out the resistance quickly to avoid excessive power dissipation. The main problem with this approach is that the introduction of a resistive component into the vector diagram of the capacitor motor (refer back to figure 3.5) will cause both current vectors to rotate back towards the voltage vector, thus affecting the phase relationship adversely. The combination of this with the voltage reduction leads to a serious reduction in torque, far greater than proportional to the current reduction. Thus this is a bad method in theory. However, it is very cheap to implement, as it requires only a substantially rated resistor and an automatic switch to short it out. The effect of such a procedure on the starting current of a three-horsepower semihermetic compressor with a single-phase motor is shown in figure 3.8. Similar results have been reported for a one-horsepower machine by

Winkler and Schneider (1979) who obtained a reduction of starting current from 30 to 9.5 A. In the experience of the authors it is quite satisfactory in practice, provided (and this is a most important proviso) that the motor is completely unloaded during the starting period. Methods for doing so will be considered in Chapter 4.

(c) Series–parallel switching of motor windings can be arranged. This usually requires special adaptation of the motor, and thus is not applicable in most circumstances. This is a pity, because it is by far the most satisfactory arrangement from the viewpoint of minimising the torque reduction. It is to be hoped that compressor manufacturers will bear this in mind in future designs.

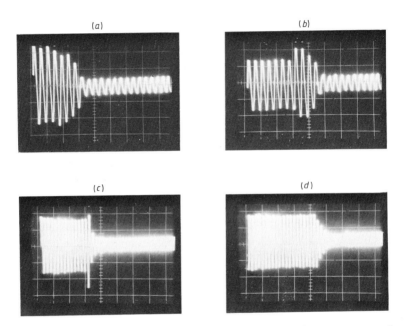

Figure 3.8 Starting current of a single-phase three-horsepower semi-hermetic compressor. Vertical scale 50 A (peak) per major division. (*a*) Direct on-line start. Peak current = 110 A. (*b*) With 1.5 Ω in series, shorted out after about 7 mains cycles (140 ms). Note the significant number of cycles at 100 A, with the initial part of the current reduced by the resistor to a more acceptable 70 A. (*c*) Resistor shorted out after about 15 mains cycles (300 ms). Only about one cycle is now at 100 A. (*d*) Resistor shorted out after about 20 mains cycles (400 ms). There is now no high-current region, because the starting capacitor has been disconnected just before the series resistor is shorted out. Obviously (*d*) is the most desirable characteristic, though the starting period is rather longer than one might like.

In all cases it is necessary to arrange a switchover from 'starting' to 'running' conditions. Theoretically this is best done by sensing the speed of rotation, but a timer is simpler and cheaper. In the authors' experience a period of about one second is generally enough for starting most domestic-sized compressors provided they are off-loaded during start-up. This is such a short time that even if the compressor should completely refuse to start until 'running' voltage is applied, neither the motor nor the starting capacitor is likely to come to much harm. This fail-safe feature is an added reason for preferring a timer to a speed-sensing method of operation; see Chapter 4.

3.4 Motor speed control

As will be discussed in Chapter 4, it is desirable to vary the speed of the compressor so as to match the output of the heat pump to the needs of the load. This can be difficult with electric drives, since induction motors are inherently single-speed machines. Obviously it is possible to reduce the speed by increasing the slip, and this can be achieved simply by reducing the voltage. However, the efficiency is seriously reduced under conditions of high slip, and it is therefore an undesirable method of speed control for our purposes.

An alternative approach, which yields a series of fixed speeds rather than a continuously variable speed, is to use a switchable stator winding. A common arrangement is to switch from two poles to four, thus halving the speed of rotation. This is a highly efficient means of speed control as it involves no increase in slip. However, it may be objected that halving the speed, though highly convenient for start-up, may be an excessive amount of speed reduction for many purposes. It is obviously difficult to arrange for a small reduction in speed by means of a small increase in the number of poles, unless the number of poles is rather large, which implies a rather low speed of rotation, and thus is unlikely for compressor applications. To circumvent the problem, it is possible to generate effective poles which can be moved around by appropriate modulation, giving a speed reduction which is considerably less than that produced by pole switching, but such techniques do not appear to have become commonplace and thus are unlikely to be useful to the heat pump designer in the immediate future.

The most satisfactory method of speed control, which has the advantage of being infinitely variable rather than varying in a series of steps, is to vary the frequency of the supply by electronic means. This allows the slip to remain small, and thus the efficiency to remain high, despite the speed reduction. For this type of supply it is probably preferable to design special motors, and the reader is referred to de Jong (1976) for details. However, it is perfectly possible to use conventional motors, and the designer will in most cases be constrained to do so because of price and availability.

An electronic converter to produce the required variable-frequency supply involves rectification of the AC power from the line, followed by a certain amount of smoothing to produce a reasonably constant DC, and then a network of controlled rectifiers to chop up the DC into something resembling conventional polyphase AC. There is, of course, no particular advantage in generating single-phase AC rather than three-phase by such means, because one would then be in the disadvantageous position of having to use single-phase motors. Thus in circumstances where only a single-phase mains supply is available, an electronic converter has the dual advantage of variable speed and of supplying three-phase power. Nevertheless, there is still a disadvantage in single-phase mains supplies, because the amount of smoothing needed to obtain a reasonably smooth DC is much greater than in the case of a three-phase supply.

The converter can be of either voltage-source or current-source type, although some may be a mixture of both. In the former, a well smoothed DC voltage is chopped into a staircase approximation to a sine wave, as shown in figure 3.9. This curiously shaped waveform does not appear to worry the motor, the current in which is smoothed by the inductance of the motor coils to a remarkably sinusoidal shape. The main problem, which is acute when the source is single phase, is that there are inevitably times when the supply voltage is near zero, and unless a reserve of energy is provided by a capacitor, there may be a shortfall of power at a time when the output requires it. The size of the capacitor needed to overcome the problem is formidable; in fact it probably represents the major part of the cost of the whole system. If a three-phase supply is available, the smoothing problem is of course less severe. A current-source converter, in which a substantial series impedance forms the major current limiting factor in the circuit, further minimises the smoothing problem, and if the impedance of the series choke is high enough, it may be possible to dispense with a capacitor completely, though probably not in the case of single-phase power sources. In the current-source converter the current is transferred from one phase to another in turn, leading to a curious current waveform, but again this does not appear to worry the motor.

As an alternative to controlled rectifiers, it is possible to use transistors, especially in smaller power converters. Transistors are more expensive, but they can switch themselves off when required, unlike controlled rectifiers which have to wait until the current falls to zero before they cease to conduct. The circuitry to drive the power stages can therefore be rather simpler.

Converters are available commercially, and appear to work very well (see for example Ek 1978) but they are usually intended for larger motors from about 10 kW up. However, the economics of controlled rectifiers are steadily improving, and it is likely that converters for small motors will soon be worth consideration.

Several useful side effects of the use of converter supplies of variable frequency are worthy of note, because they overcome some of the serious problems of heat pumps in domestic installations. The first and most important is that because the motor can be run up relatively gently from rest, starting currents are reduced. This is assisted by the requirement that in order to maintain the same magnetic flux, the voltage should be proportional to the frequency (Ek 1978); this automatically implies a low starting current in the motor, and thus in the mains supply to the converter. A second useful feature is that the reactive component of the motor current, which, as we have seen already, can be quite large at low loads, can be confined to the circuit from the converter to the motor, while the power drawn from the mains can be much closer to being purely resistive, with a much improved power factor. Thirdly, according to de Jong (1976) there are certain disadvantages in two-pole induction motors which lead to lower efficiency, and when a converter supply is used it is no longer necessary to use two-pole motors for high-speed work, since one can increase the frequency and use four-pole motors. Finally, and related to the previous comment, the possibility of higher-frequency power may allow the use of physically smaller motors, which may be advantageous in some circumstances.

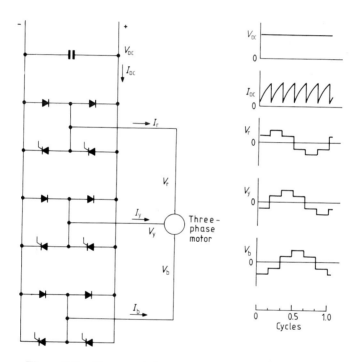

Figure 3.9 Characteristics of a voltage-source converter.

There are several other types of electric motors in which speed control is inherently easier than in squirrel-cage motors. The wound-armature induction motor has an external resistance, connected in series with the armature via slip rings, which can be varied to control the amount of current in the armature, and thus vary the slip. As with any variable-slip system, the efficiency is poor at high slip, so this is an unattractive method of control for our purposes, but there is the added serious disadvantage that slip rings and brushes are needed, which have obvious maintenance requirements and are quite unsuitable for hermetic compressors. Commutator motors, whether DC or AC, are also unattractive on grounds of brush maintenance, which is a pity because otherwise they are attractive for speed control.

3.5 Variable-ratio drives

As an alternative to variable motor speed, it is possible to vary the gear ratio between motor and compressor, leaving the motor to look after itself. This may appear to be a rather crude approach, but at least it can be made to work at reasonable cost—a matter of no small importance.

The most sophisticated method of variable-ratio drive is probably hydraulic transmission. It is, of course, well established in vehicle and railway locomotive propulsion. It has also been used in very small power capacities in such items as garden tractors, but in such cases it is chosen on grounds of operating convenience rather than efficiency, and it must be admitted that it probably wastes a considerable amount of power. Nevertheless, the Central Policy Review Staff (1974) proposed the use of variable-speed hydraulic transmission in factories, and estimated the resulting fuel saving for the UK as about 0.3 million tons of coal equivalent per year, so it does seem to offer some potential advantages for heat pump drives. The main disadvantage is probably cost.

A cheaper alternative, and one which offers a high efficiency, is the variable V-belt drive, familiar from its use in a small car of Dutch origin. Belts are notoriously unreliable, but if they are adequately serviced it should be possible to get a lifetime of considerably more than one heating season, which should make them acceptable for most purposes. The most serious objection to belt drive is the impossibility of using a hermetic or semihermetic compressor.

Instead of infinitely variable systems, it may be worth considering a system with a number of fixed gear ratios. Whether in the simplest form of stepped pulleys, or in a more complex form such as an automobile gear box, this is inevitably less easy to control automatically than infinitely variable arrangements, so it is less suitable for the present purpose. Nevertheless, the possibility of a manually changed two-speed system, with low compressor speed for autumn and spring, and a higher speed for winter, should not be overlooked.

As noted earlier, the part-load operation of an electric motor tends to be less efficient than full-load running, and this must be taken into account in any variable-ratio drive. It should also be remembered (and this applies to motor speed control as well) that there may be a degradation of compressor efficiency when the speed is significantly different from the design speed (see §4.4.1). Care must be taken that any improvements in heat pump performance resulting from better load matching due to speed control are not lost by the reduction in efficiency due to these two effects.

3.6 Fuel-fired compression heat pumps

The thermodynamic advantage of fuel-fired heat pumps, as pointed out earlier, is that the waste heat rejected by the low-temperature end of the engine is available to supply useful heat rather than, for example, going to warm the sea around a power station. There is an added attraction in that the temperature of the waste heat can be and usually is higher than the condensing temperature of the heat pump, so it can be used via a heat exchanger to raise the temperature of the heat supplied by the heat pump. Since a principal objection to heat pumps in many applications is the low temperature of the heat supplied, this is an advantage which should not be neglected.

A typical heat flow diagram for such a machine is shown in figure 3.10. Most of the waste heat from the engine enters the place where it is needed, some of it coming from the cooling water and some (probably the majority) coming from a heat exchanger in the exhaust pipe. Some heat is lost by convection etc from the warm machinery, and a rather larger amount is lost from the exhaust.

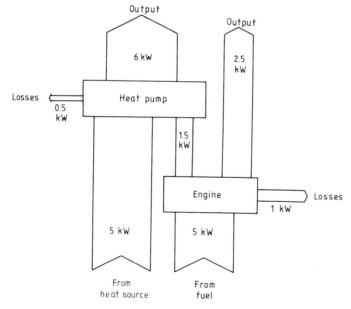

Figure 3.10 Heat flow diagram of an engine-powered heat pump. A COP of 4.0 has been assumed for the heat pump, a mechanical efficiency of 30% has been assumed for the engine, and the engine waste heat is taken to be collected with an efficiency of 70%. All of these assumptions are optimistic.

It is worth pointing out that it is considerably easier to make use of the waste heat from an external combustion, closed-cycle engine than from an internal combustion, open-cycle engine, so the conventional petrol-, diesel- or gas-fuelled engine is at a disadvantage compared with Rankine (steam) engines, Stirling engines etc. This is in complete contrast to the situation in automotive engines, where the need for a heat exchanger to reject large amounts of waste heat to the atmosphere is a serious problem.

It is possible to consider all manner of engines for the job of driving the compressor. As well as the obvious petrol- (gasoline-) fuelled reciprocating engine, it is possible to consider a very similar engine fuelled with natural gas. We can also consider rotary engines such as the Wankel, the gas turbine (otherwise known as the Brayton-cycle), and of course the ubiquitous and highly efficient diesel engine. Among Rankine-cycle engines it is possible to look at not only steam engines and turbines, but also at vapour driven machines using other working fluids, including the Freon refrigerants. Finally it is possible to consider several types of Stirling cycles, most of which are at best experimental and at worst mere gleams in the eyes of their inventors. One important point to bear in mind is that domestic-sized heat pumps require rather small sizes of engine, because the condenser heat output is supplemented by the heat from the engine. This can actually be a disadvantage because there is a shortage of really efficient small engines; the thermodynamic efficiency of a lawn mower is not usually of prime importance! On an industrial scale, however, there is no shortage of suitable engines.

An overall comparison of the possible utility of different types of engines in the United States has been made by Dutram and Sarkes (1979). They attempted to estimate the cost saving per 'ton', compared with a conventional gas furnace, for heat pumps of different types fired by natural gas. While this is obviously sensitive to fuel costs as well as to the capital cost and performance of the systems chosen for comparison, the results were sufficiently consistent for all climatic regions of the USA to make Stirling cycles seem very attractive. On the basis of this comparison, the General Electric group has proceeded to develop a Stirling-engined heat pump using a free-piston engine with helium at 60 atm pressure as working fluid, and a top temperature of 650 °C, for which the predicted efficiency is as high as 30–35%, while the burner which supplies heat to the engine has achieved efficiencies of 85%. The engine drives an inertial compressor in which the cylinder is moved to and fro while the piston remains stationary, thus eliminating sealing problems, and an isentropic efficiency of 85% is expected.

An even more radical departure has been adopted by Strong and co-workers, who use a Rankine power cycle driving a Rankine refrigeration cycle. Notwithstanding the unfavourable position allocated to a Rankine/Rankine configuration by Dutram and Sarkes, the comparison produced by Strong (1979) came down heavily in favour of such a device. The key factor is the development of a power cycle in which the working fluid is one of the Freon refrigerants, and the power converter is a tiny high-speed turbine rotating at speeds of the order of 100 000 rpm, with an equally tiny centrifugal compressor at the other end of the shaft. At such high speeds it is necessary to use gas bearings in which the metal surfaces are kept apart by a stream of the working fluid. One problem with such a machine is the need to use the same Freon for both turbine and refrigeration cycles, with an inevitable compromise in thermodynamic performance. Higher overall efficiencies could be obtained if different fluids could be used for the two duties, but this would necessitate a shaft seal between the turbine and the compressor, which at 100 000 rpm is no small problem.

The most straightforward approach to engine-powered heat pumps is to use a conventional petrol (gasoline) or diesel engine. Extraction of heat from the coolant is straightforward, especially if the engine is water cooled. Extraction from the exhaust gases, a considerably larger source of heat than the coolant, is less straightforward because of the danger of condensation, with consequent corrosion, if the gases are cooled below dew point. Nevertheless, it is desirable to cool the gases as far as possible to promote good efficiency, and it may be worth using stainless steel heat exchangers and special draining arrangements to allow the condensate to run away harmlessly. Incidentally, it is customary to quote net calorific value when specifying the energy content of a fuel, and if one condenses some of the combustion products, one goes some way towards extracting the rather larger amount of energy represented by the gross calorific value. Thus one can apparently obtain a fuel efficiency of more than 100% of the conventional net CV even before one begins to drive the heat pump.

A number of engine-driven heat pumps are in use in Germany (van Heyden and Wölting 1979). These are mostly rather large machines, typified by an open-air swimming pool installation at Dortmund, with an output of about 700 kW, driven by an engine of about 80 kW mechanical output, running on the perhaps unexpected fuel of ordinary town gas rather than natural gas.

One particularly useful feature of engine drive is the ease with which the speed can be varied to suit the needs of the load. This is particularly true of the diesel engine, which has a rather flat curve of efficiency, or alternatively of specific fuel consumption in g/kWh of output, against speed. A typical curve for a small four-stroke diesel engine is shown in figure 3.11. The efficiency falls off at low power output, mainly because of frictional losses which become more significant at lower outputs, and also at high power when the combustion becomes less efficient because the fuel–air ratio becomes higher than the optimum when the engine is being overworked. A petrol engine is somewhat less suited to variable-speed duty because of its different fuel induction system, but it is nevertheless useable over a useful range of speed. The characteristics of a heat pump driven at variable speed by a petrol engine have been studied in a computer model by Fleming (1978). Some of the results of this study are shown in figure 3.12, and the effects of the reduced compressor speed are obvious. Thus in milder winter weather when the reduced output at low-speed operation is sufficient to satisfy the heat demand, the benefits of speed control may be considerable.

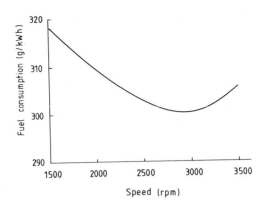

Figure 3.11 Typical fuel consumption characteristic for a small diesel engine. Maximum efficiency is taken to be 300 g/kWh mechanical output; this corresponds to about 25% efficiency (referred to gross CV of fuel).

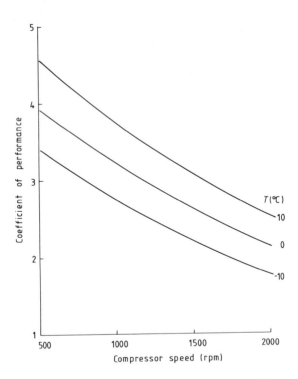

Figure 3.12 Computer simulation of a variable-speed heat pump for different temperatures of the source air. The improved COP at low speeds results from the smaller temperature differences across the heat exchangers. (Reproduced by courtesy of H Fleming.)

One of the driving forces towards engine-driven heat pumps is the availability and cheapness of natural gas in some parts of the world. This is not an appropriate place to discuss the long-term future of natural gas supplies, but it is appropriate to mention one of the difficulties that may arise in running conventional automotive-type engines on such a fuel. Because of the complete absence of tetra-ethyl-lead (antiknock additive) in the fuel, there may be some lack of lubrication. According to Pegley and Rieke (1979), the main problem is excessive wear of the valve gear, and it can be substantially overcome by preventing rotation of the valve relative to its seating.

Whether gas- or liquid-fuelled, the whole question of the useful life of an engine-powered heat pump is rather problematic. Critics point out that if one imagines a typical annual duty of about six months of heating season with the machine running for about 50% of the time, this represents an annual running time equivalent to about 100 000 miles of car travelling, at which point one usually contemplates changing the engine for a new one, if one has not already sold the car (and probably also the one that was bought afterwards!). However, according to the proponents, it is possible to extend the interval between major services to the equivalent of a complete heating season, though it is sometimes by no means clear what is meant by a major service in this context.

An intriguing way round the problem of engine lifetime has been tried by Rummel (1979) who set up a heat pump with both an electric motor and a diesel engine as drives. The electric motor was used in milder winter conditions, changing over to the engine (with its waste heat) for more extreme conditions. This undoubtedly has some advantages, and may seem to offer the best of both worlds, but the cost penalty seems likely to be a significant deterrent to its widespread adoption.

4

CONTROLS

The control of a heat pump can involve anything from a simple thermostat to a complicated computer-based system. There is often no easily defined way of choosing a suitable level of sophistication for a particular application, but since it is almost inevitable that a more complicated control system will cost more than a simple one, economic considerations may well be of greater importance than technical ones in choosing what to control and how. Because of this problem, the present chapter will consider a number of individual aspects of control, without necessarily implying that all of them might usefully be incorporated into a design.

4.1 Safety devices

A heat pump operates under higher temperatures and pressures than the corresponding refrigeration system, and it is more liable as a consequence to damage itself and even to become hazardous. The incorporation of suitable safety devices is of considerable importance, therefore, but because they cost money and do not usually contribute to a higher coefficient of performance, they should not be overelaborate or excessively redundant.

Normal refrigeration practice makes use of a high- and a low-pressure cut-out. An excessively high pressure in the discharge from the compressor is usually the result of insufficient heat dissipation from the condenser. This may or may not be a permanent situation, so the high-pressure cut-out is usually made selfresetting. An excessively low pressure at the compressor inlet is usually the result of loss of refrigerant. This is potentially damaging to the compressor, because it may cause air to be sucked into the refrigerant circuit, and may also prevent oil from circulating correctly, and so it is customary to make the low-pressure cut-out non-selfresetting. In heat pumps the same strategy can be adopted. However, because it is possible for a low evaporator pressure to result from a very cold day, rather than from loss of refrigerant, it is necessary to take care with the adjustment of the low-pressure cut-out, especially if it is made non-selfresetting.

Overheating cut-outs are highly desirable additions to most electric motors, and are very cheap. Occasionally compressors have overheating sensors, usually thermistors, to monitor the temperature of likely trouble-spots such as the cylinder head (or the equivalent part of the anatomy for a rotary compressor) and the discharge port. Although discharge temperature and condenser pressure are related, it is probably not safe to rely solely on temperature or pressure sensors for protection against malfunction of the compressor and condenser, because an excessive amount of superheating may cause very high temperatures without a corresponding pressure excursion, as might a shortage of lubrication. Also, a temperature sensor is likely to be much slower in operation than a pressure trip, which could respond to an excess in one single stroke of a compressor piston. Further, the pressure of a boiling liquid is exponentially related to the temperature (see figure 2.15), so that pressure-operated switches provide an inherently more sensitive means of protection. On the other hand, it is possible to use temperature sensors with an output exponentially related to temperature, so that temperature-operated systems are not necessarily much less sensitive than pressure systems.

A rather more subtle form of overheating protection can be incorporated as part of a capacity modulation system, as will be described in § 4.4. At the other end of the scale of subtlety, the use of fusible plugs or pressure release valves in condenser vessels is a cheap way of insuring against disaster, although they have been known to leak with consequent loss of refrigerant. A slightly more satisfactory version of a pressure release valve is incorporated in some types of compressor in the shape of a spring-loaded head plate which will be pushed back by excessive pressures, opening up a pathway from the high- to the low-pressure side. Since this does not vent the refrigerant to air, a certain amount of leakage can be tolerated in such a device.

4.2 Defrosting

The formation of ice on the evaporator of an air-source heat pump is quite inevitable whenever the evaporator temperature is below freezing point unless the source air is exceedingly dry, and if a thick layer of ice is allowed to build up, the heat transfer capacity of the evaporator is bound to be diminished, so the ice must be removed when necessary. However, it should not be assumed that the energy required for de-icing is wholly a loss to the heat pump system. The energy required to melt (and thus, one hopes, remove) one gram of ice at 0°C is 325 J, but the energy acquired by the evaporator in condensing one gram of water vapour to liquid at 0°C is 2500 J, together with a further 325 J on freezing the liquid water to solid ice. If one assumes that the defrosting energy is applied with perfect efficiency, the net energy gain to the evaporator is exactly equal to the enthalpy of condensation. This is so much greater than the energy involved in defrosting that it would require an extremely inefficient defrosting system before the net energy gain to the system became negative.

The refrigeration industry has developed several methods for removing the ice periodically. The simplest method, applicable only when the source air can be relied on to be appreciably warmer than 0 °C, is simply to stop the compressor and to continue to blow air over the evaporator coils. This is unlikely to be useable in most heat pump installations. A variant of this method is to use heated air, which might be available, for example, from a ventilation exhaust duct in a commercial building. A method of more universal application, and one which is widely used in refrigeration of cold-rooms, is to incorporate electric heater elements into the coil block, either within the finned space of a fin-and-tube construction, or within the tubes themselves. The latter is more efficient, since there is less loss of heat to the environment, but the former is easier to make. Yet another cold-room technique is to spray the evaporator with water, or even brine. This is doubtless effective, since the transfer of heat to the frost layer is likely to be very good, but it is unlikely to be appropriate in the majority of heat pump installations.

The neatest method is to use the refrigerant itself as the heating medium. This is common in reversible heat pump/air conditioning machines, in which a reversing valve is used to interchange the roles of evaporator and condenser. When defrosting of the heat pump evaporator is needed, the machine is reversed, thus acting as an air-conditioner for a few minutes, withdrawing heat from the warm side of the cycle and feeding it into the evaporator. This can be extremely effective, and so rapid that the heated space suffers no serious cooling; even if it does, it is possible to bring in electric resistive heating or some other short-term source of heat during the defrosting cycle. In most cases it is necessary to use a second expansion valve to control the flow of refrigerant into the condenser during its brief activity as an evaporator, and some ingenious methods have been devised for optimisation (Trask 1979). The main criticism of reversing as a method of defrosting arises in heating-only heat pumps where there is no need for an air conditioning ability, because one is then introducing a fairly expensive and somewhat unreliable item, the reversing valve, into a system which otherwise has no need for it. A second criticism is the ability of the reversing valve to act as a remarkably good heat exchanger, not necessarily in a part of the cycle where one would wish to use it, because it raises the superheat of the gas entering the compressor, leading to a higher discharge superheat.

Another way of using the refrigerant as the heating medium is to pass the hot gas directly from the compressor outlet into the evaporator. This can be achieved by a simple magnetic valve, bypassing the expansion valve and the condenser, and is thus simpler to implement than a reversing system. However, it is of limited defrosting capacity, because once the initial pressure has been dissipated, the compressor acts merely as a pump, working against a very small pressure difference. Consequently if this method of defrosting is to be useable, a good method of sensing the presence of frost is essential. A further disadvantage of the method is the possibility of liquid entering the compressor because the expansion valve is not able to control the flow. In practice this can usually be minimised by careful layout of the pipe runs. It is worth remarking that both the disadvantages just mentioned are of much less consequence with rotary sliding-vane compressors; such machines are tolerant of reasonable amounts of liquid, and because of the effectively fixed volumetric compression ratio, there is always a certain amount of compression of the refrigerant even if the bypass valve is open, so a continuous defrosting operation is possible. On the whole this is a fairly good method of defrosting, if used with discretion. The initial defrosting effect is extremely effective, and the authors have found it satisfactory in both severe and mild winter conditions. It has the interesting advantage that the magnetic valve used for bypassing the hot gas also serves to equalise pressures and unload the compressor during start-up, thus obtaining two useful functions from one fairly cheap component. This can also ensure that a defrosting operation occurs every time the machine stops, which may be slightly extravagant in energy but can be very effective in minimising the number of times a 'real' defrosting cycle is needed.

An alternative to the thermal methods described above is the mechanical method proposed by Richardson and Husker (1975). A set of scrapers serve both as a fan and as a means for removing ice, and if the spacing is sufficiently small, the remaining layer can be kept thin enough to have a negligible effect on heat transfer. An interesting side-effect is that close spacing between the blades and the heat exchanger seems to break down the boundary layer and to improve the heat transfer coefficient. Early versions of Richardson and Husker's device suffered from mechanical difficulties and noise, but it is nonetheless worthy of further investigation.

4.3 Methods of initiating defrosting

With the exception of the mechanical system just described, all the defrosting methods rely on intermittent operation, and although they may be satisfactory in themselves, they can only be fully effective if they are operated at a time when the ice build-up on the surface is great enough to interfere with the transfer of heat, but not so great that it cannot be removed easily. Also they must desist when the ice has been removed, but only just so. There are several methods of initiating a defrost cycle, each of which has its own advantages and difficulties.

4.3.1 *Time-and-temperature method*
This method works on the assumption that if the temperature of the evaporator (or of the source air) falls to a certain level below freezing point, a defrosting cycle of a fixed period will be needed after a certain well defined period of operation. Thus in essence we need a temperature-operated switch and two timers, one for defrost period and one for running period. The temperature switch can be replaced by a refrigerant pressure switch on the evaporator, because evaporator pressure and temperature are related. Furthermore, it is possible to dispense with the defrost period timer if one assumes that whenever the evaporator rises

above a preset temperature or pressure, defrosting is complete. This leads to an extremely simple defrost controller (Trask 1979) which is sketched in figure 4.1. However, the intitial assumption is somewhat questionable; a low evaporator temperature could well result from source air which is extremely cold *and dry*, causing very little frosting, and there is no reason to assume that the preset period is an optimum one. Since in practice it is necessary to ensure adequate defrosting under the worst circumstances, a simple controller of this type will usually err on the side of excessive defrosting, with a significant number of unnecessary defrost cycles, unless the climate in which it is used is highly predictable. According to Pollard (1975) the actual consumption of energy for defrosting, as a percentage of the total electricity consumption of the whole system, is quite small (2.4% in the trial reported by Pollard), so it would seem to be permissible to tolerate some spurious defrosting cycles. However, because it is not uncommon to use electrical resistive heating as a standby heat source for the building during defrost cycles, the overall increase in electricity consumption due to defrosting may be larger, and thus it is desirable to avoid too many spurious cycles and to limit the duration of defrosting operations to the minimum necessary to ensure complete frost removal. In the variable climate of the British Isles and the coast of northern Europe it seems unlikely that a time-and-temperature method can ever be optimised, and there will inevitably be numerous spurious defrosting cycles (Heap 1977), but it does seem that the method can be made to work effectively and reliably.

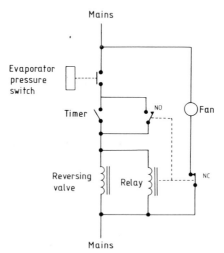

Figure 4.1 Time-and-temperature defrost controller. If the evaporator pressure is lower than the preset switching point, then when the timer contacts close, the reversing valve operates and defrosting begins. The relay also operates, stopping the fan and also shorting out the timer contacts, thus latching the controller in the defrost mode until the evaporator pressure reaches the preset upper switching point. NO, normally open; NC, normally closed.

4.3.2 *Thermal resistance detection*

When ice builds up on an air-source evaporator, the difference in temperature between the evaporator coils and the source air is increased because of the thermal resistance of the ice layer. A pair of temperature sensors and a comparator amplifier can therefore be used as the basis of an ice detector (Buick *et al* 1978).

The most straightforward way of temperature measurement for the present purpose is by change of electrical resistance and the sensors can be either thermistors or platinum resistance thermometers. Either of these types of sensor can be used in similar circuit arrangements, the simplest of which are the potential-divider arrangements (figure 4.2), and the constant-current arrangement (figure 4.3).

Let us define the relationship between the resistance of the sensor and its temperature by

$$Z = f(T)$$

and assume both sensors have identical characteristics. Then for the potential-divider circuit

$$V_a = V_s Z_a / (R_a + Z_a),$$
$$V_c = V_s Z_c / (R_c + Z_c),$$

and for the constant-current circuit

$$V_a = I_a Z_a,$$
$$V_c = I_c Z_c.$$

In both cases the value of interest to the comparator is

$$\Delta V = V_a - V_c.$$

The manner in which this varies with temperatures T_a and T_c and with temperature difference $\Delta T = T_a - T_c$ will depend not only on the type of circuit but also on the relationship $Z = f(T)$. We must now examine this point for the types of sensor likely to be used in practice.

Thermistors are semiconductor devices, and consequently have a logarithmic temperature–resistance characteristic. It is relatively straightfor-

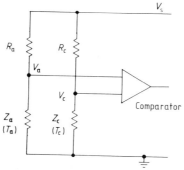

Figure 4.2 Potential-divider circuit.

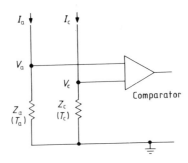

Figure 4.3 Constant-current circuit.

ward to fit a suitable equation to the manufacturers' published performance graphs, and having done so it becomes easy to compute the voltages V_a and V_c across the thermistors for different temperatures T_a and T_c and for known fixed values of fixed resistors R_a and R_c (for the potential-divider arrangement) or for known fixed values of constant current I_a and I_c (for the constant-current arrangement). It is easier and perhaps more obvious to calculate ΔV for a fixed value of ΔT, but in a real circuit it is likely that a comparator will be set to change its output state, and thus initiate defrost, when a certain value of ΔV is reached.

Let us calculate, therefore, the value of ΔT which is necessary to generate a certain fixed value of ΔV for a series of different air temperatures, T_a, all other circuit parameters being known and constant. The result of such computations, for both potential-divider and constant-current circuits, is shown in table 4.1. It is clear that for a potential-divider connection, the value of ΔT required to initiate defrosting is greater at lower temperatures, while for the constant-current connection it is greater at higher temperatures.

What is the effect of the above on a real defrosting system? The potential-divider circuit increases the chance of a defrost cycle at high temperatures; therefore it is likely to cause unnecessary defrosting cycles at high air temperatures, and may not generate adequate numbers of defrosting cycles at low air temperatures. On the other hand, the constant-current circuit decreases the chance of a defrost cycle at high temperatures; therefore it tends to prevent unnecessary defrosting cycles at high air temperatures, but also to risk excessive numbers of defrost cycles at low air temperatures. Of the two, the constant-current connection is definitely to be preferred. In practice it is unlikely that the apparent problem at low air temperatures will arise, for two reasons. Firstly, the rate of heat extraction is reduced at low air temperatures, and so the value of ΔT during operation is reduced, so lessening the risk of a spurious defrost cycle. Secondly, with any constant-current source there is a limit to how high the voltages across the thermistors can rise—usually this limit is a little less than the supply voltage.

Table 4.1 Performance of different circuits and sensors.

| Air temperature T_a (°C) | Temperature difference ΔT to initiate defrosting (°C) | | | |
| | Thermistor | | PRT | |
	Potential divider	Constant current	Potential divider	Constant current
−20	21.9	3.4	7.7	8.5
−10	12.1	6.5	8.0	8.5
0	9.3	10.8	8.3	8.5
10	8.3	16.3	8.7	8.5
20	8.2	23.0	9.1	8.6

Notes

Thermistor: GL23.
Potential-divider circuit: $R_a = R_c = 2\,k\Omega$, $V_s = 12\,V$, $\Delta V = -1\,V$ ($V_c > V_a$)
Constant-current circuit: $I_a = I_c = 1\,mA$, $\Delta V = -4\,V$ ($V_c > V_a$)

Platinum resistance thermometer: Pt-100
Potential-divider circuit: $R_a = R_c = 100\,\Omega$, $V_s = 12\,V$, $\Delta V = 0.1\,V$ ($V_a > V_c$)
Constant-current circuit: $I_a = I_c = 30\,mA$, $V = 0.1\,V$ ($V_a > V_c$)

So far we have treated the potential-divider circuit and the constant-current circuit as if they were entirely distinct arrangements. However, a common practical arrangement for getting an almost constant current through a variable resistance is to supply current from a high-voltage source via a high fixed resistance whose value greatly exceeds the largest expected value of the variable resistance. We can now see that a constant-current circuit is in fact a limit of the potential-divider circuit, with V_s very large and with R_a and R_c both very high. Since we have found that a potential-divider circuit (with small R_a and R_c) tends to increase the chance of defrosting at high temperatures, and that a constant-current circuit (with very high R_a and R_c) does the opposite, it is reasonable to assume that there is a range of values of R_a and R_c in which neither of these behaviour patterns dominates. If the computations are repeated for increasing values of R_a and R_c using the potential-divider formulae, we find that this is indeed the case, as is shown in table 4.2. Further, this mixed behaviour is reached at an eminently reasonable value of R_a and R_c and with reasonable ranges of voltages across the thermistors; in other words, this state is attainable.

An alternative electrical temperature sensor is the platinum resistance thermometer, which in contrast to a thermistor, has an almost linear relationship between resistance and temperature, and thus is inherently more suitable for our present purposes. Until recently the cost of such devices was rather high, but the introduction of thin-film thermometers at a price comparable to precision thermistors makes them worth considering.

It is easy to fit an equation to the manufacturers' published performance data, especially as they provide precise tables in exactly the required form. Computations for a potential-divider circuit and for a constant-current circuit are shown in table 4.1. Constant current is clearly better for obtaining a constant ΔV and ΔT, but a potential divider is not much worse and the deviation is in the 'preferred direction'. The difference between this and the thermistor calculations arises partly from the linear behaviour and partly from the relative insensitivity of the resistance to temperature.

Table 4.2 Effect of increasing fixed resistors in a potential-divider circuit.

R_a (kΩ)	ΔV (V)	Value of ΔT at stated ΔV for different T_a (in K)				
		-20	-10	0	10	20
2	-1.0	21.9	12.1	9.3	8.3	8.2
3	-1.3	19.1	12.0	10.0	9.7	10.4
4	-1.3	13.8	10.0	9.1	9.5	10.7
5	-1.4	12.4	9.7	9.4	10.2	12.0
6	-1.4	10.8	9.1	9.2	10.4	12.6

Notes

GL23 thermistor, $R_a = R_c$, $V_s = 12$ V. ΔV chosen to give a reasonable range of ΔT.

The value of ΔV in table 4.1 is rather small, and this might create problems, especially if there is significant hum or other noise present. The only way of increasing ΔV is to increase the current through the sensors, and unfortunately this is limited by selfheating. A further disadvantage of platinum resistance thermometers, at least in their low-cost thin-film form, is their sensitivity to mechanical damage. Thermistors, being smaller, tend to be more robust. One possible advantage of platinum thermometers is their greater precision of manufacture, leading to closer similarity of resistance–temperature behaviour between one example and another of nominally identical type. However, Buick *et al* (1978) have shown that even with non-precision thermistors, the effect of manufacturing tolerances can be rendered negligible in practical circuits.

Instead of resistance thermometers, it is possible to use a pair of temperature-sensitive integrated circuits. Several different types are available, producing either a current of about 1 μA or a voltage of about 1 mV per kelvin. Such devices are almost exactly linear in their response to temperature, which is very convenient for the present purpose. They are able to operate over a wide temperature range, typically −50 to 150°C. Their shape and size makes them less easy to use than thermistors, though probably no more difficult than platinum resistance sensors, and their cost, though higher than either thermistors or platinum thermometers, is not unacceptable.

There is a basic problem with any temperature-difference sensing ice detection system in that when defrosting, the temperature difference between the evaporator coil and the air will be reduced, and perhaps even inverted by the action of the defrosting system. In other words, the device tends to switch itself off! To prevent this, it is necessary to make the ice detector operate a timing circuit, and to set the time period to some suitable value depending on how long the defrosting system takes to remove the ice. A practical circuit, using platinum resistance sensors, is shown in figure 4.4. It has been found to work well in a wide range of conditions over a period of two years, and can be regarded as satisfactory.

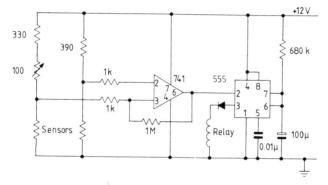

Figure 4.4 Practical circuit using Pt-100 platinum resistance thermometers as sensors. A small amount of positive feedback is applied to the 741 amplifier to encourage positive switching.

4.3.3 *Capacitance detection*

Of the physical phenomena which can be used for ice detection, the dielectric constant is one of the most useful. At low frequencies the dielectric constant of water is about 79, and that of ice is about 75, so it is not possible to distinguish between the two by this parameter alone. On an evaporator coil, however, ice can build up to a considerable thickness whereas water can be made to run away, so it is possible to distinguish between them on the basis of the thickness of a dielectric between the plates of a capacitor (Buick *et al* 1978). This is aided by the fact that thin layers of dielectric have much less effect on the capacitance of a capacitor than thick layers do. To see that this is so, let us consider how the capacitance of a pair of parallel plates behaves when the dielectric consists of a layer of air and a layer of another medium of high dielectric constant, and the interface between the dielectrics is allowed to move.

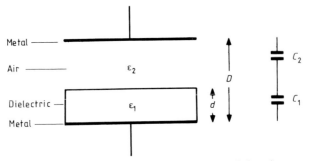

Figure 4.5 Capacitor with ice and air dielectrics.

In the situation illustrated in figure 4.5, the total capacitance is given by

$$\frac{1}{C} = \frac{1}{C_1} + \frac{1}{C_2}$$

$$= \frac{d}{4 \pi \varepsilon_0 \varepsilon_1} + \frac{D - d}{4 \pi \varepsilon_0 \varepsilon_2}.$$

where ε_0, ε_1, and ε_2 are the dielectric permitivities of free space and each of the layers, respectively. Consequently,

$$C = 4 \pi \varepsilon_0 \frac{\varepsilon_1 \varepsilon_2}{\varepsilon_2 d + \varepsilon_1 (D - d)}.$$

Putting in numbers, we have $\varepsilon_0 = 8.85 \times 10^{-12}$ F m^{-1}, $\varepsilon_2 = 1$, $\varepsilon_1 = 75$. A typical plate spacing D might be 5 mm, and if C is computed for different values of d, we obtain the results shown in figure 4.6. The dramatic change in capacitance produced by a complete layer is now evident, and it is clear that a thin layer of water will be completely negligible in effect compared with a thick layer of ice.

The change in capacitance caused by ice build-up is most readily detected by its effects upon a capacitance bridge, as shown in figure 4.7. In order to balance such a bridge, both the reactive and resistive components must be balanced. This implies that

$$R_1/R_2 = R_3/R_4,$$

and

$$1/C_1 R_2 = 1/C_3 R_4.$$

A lossy dielectric produces an effective resistance in one arm of the bridge, and since both ice and water can be lossy, this matter requires further consideration.

The bridge may be balanced 'at air', i.e. with the plates of the capacitor clear of all ice, or 'at ice', i.e. with the plates frosted up, either fully or to some specified extent. Balancing at air has the advantage that at the balance point the dielectric is a fairly predictable substance, but the presence of water films on insulators, spacers, etc may introduce a lossy component

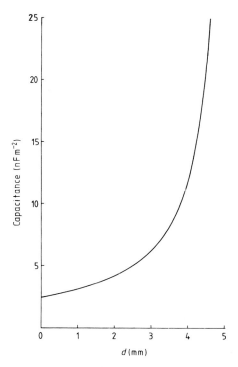

Figure 4.6 Capacitance as a function of ice build-up. $\varepsilon_1 = 75$, $\varepsilon_2 = 1$, $D = 5$ mm.

which may prevent proper balancing. Balancing at ice may not be entirely satisfactory because the ice may build up in different configurations (solid, feathery, etc) depending on the weather, and also because the setting up of a practical device would involve growing a film of ice in a realistic form before setting the adjustments; on the other hand, the problem of wet insulators does not exist.

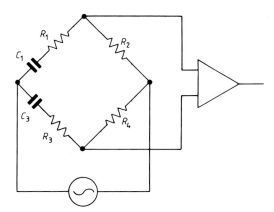

Figure 4.7 Capacitance bridge.

There are further differences between the two methods of balancing in the behaviour of the system during defrosting. This is because as the ice melts there must be a temporary increase in capacitance due to the greater dielectric constant of water in comparison to that of ice. If the bridge is balanced at air, the out of balance component increases as the ice builds up, increases further as it melts, and subsequently falls as the water runs away. This behaviour has been seen under experimental conditions. If the out of balance voltage is used as a means of operating the defrosting system, there is a built-in tendency for the system to continue defrosting until the water has run away. In contrast, if the bridge is balanced at ice, it is displaced to one side of the null (high capacitance) during melting, and to the other side of the null when the water runs away. This more complex behaviour places somewhat more difficult requirements upon the electronics, which can be satisfied to some extent by making the ice detector activate a timer.

An experimental implementation of a capacitance detection system is shown in figure 4.8. The capacitor itself consisted of two earthed outer plates, in contact with the evaporator, and an inner plate insulated electrically from the outer plates, and thus inevitably somewhat insulated thermally from the evaporator. The design of the insulators gave rise to some problems of water penetration. Balancing at air was straightforward; balancing at ice was a little more difficult to achieve.

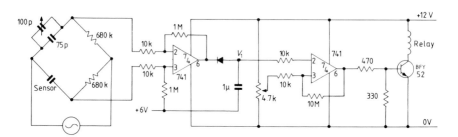

Figure 4.8 Circuit of experimental capacitance ice detector. The first amplifier functions as a precision rectifier, whose output V_1 is smoothed and applied to the second amplifier, which functions as a comparator with a small amount of positive feedback.

Initial tests indicated that the system worked, but there was some doubt about the long-term effects of water penetration into the insulators. To investigate this, a long-term test was done in which the amplified voltage V_1 was displayed on a chart recorder, together with an indication of whether the heat pump was working or defrosting. During the first few cycles, the system worked satisfactorily, but as time went on the cycles got shorter and shorter until eventually the machine 'defrosted' continuously. This is attributable to water penetration, as evidenced by the fact that the system worked again after careful drying of the capacitor assembly, and

94

also by resistance measurements across the capacitor, which showed a small but appreciable conductivity. To get round this problem, the capacitor assembly was immersed in varnish and allowed to harden before further testing. This time the system worked much better. It is still not certain, however, that the problem of water penetration has been fully solved, and so there must be significant doubts about the practical usefulness of capacitance bridges as a means of detecting ice.

4.3.4   *Air pressure detection*
If the coil block of the evaporator is blocked with ice, the fans will have difficulty in moving air through it, and a pressure difference will arise. This can be used to trigger a sensitive pressure-operated switch and thus to initiate a defrosting operation. There is a possibility of spurious triggering due to wind gusts, and some form of integrating time delay is necessary to eliminate such short-term effects. There is also the possibility of longer-term effects such as blockage of the evaporator by dead leaves, which would lead to spurious defrosting. It is probably impossible to eliminate this problem entirely, but its effect can be minimised by requiring a certain minimum air (or coil) temperature before defrosting is permitted. This would necessitate an air temperature sensor which could be used in addition as a means of varying the duration of defrosting as a function of air temperature. Such systems are available commercially as components for incorporation into heat pumps.

An amendment to such a defrost detector has been introduced in which a mechanical probe is placed within the coil block, and is free to move unless restricted by the presence of ice. If the conventional defrost controller requests a defrost cycle, the probe is energised, and if it is free to move, the defrosting cycle is inhibited. If, however, it is restricted in its movement, the defrost cycle is allowed to proceed.

4.3.5   *Radioactive methods*
Ice and water both contain large amounts of hydrogen, so it is tempting to use a neutron probe as a means of detecting the presence of an ice film. As in the case of capacitance detection, there is little difference between the values of the physical parameter for ice and water, but the ability of a solid to build up a significant thickness, and of a liquid to run away, can be used to distinguish the two states. In this case there is no problem with electrical conductivity, so it appears to be a more practicable method. The main difficulty is the cost of a suitable detector, though the question of the acceptability of a radioactive source in a domestic or commercial environment must be given serious thought as well. For both these reasons, the authors have decided not to pursue radioactive methods to any serious extent, so they are unable to comment on the usefulness of the method in practice. Nevertheless, it is a technique worth bearing in mind for the future.

4.4 Capacity modulation

As was discussed in Chapter 1, it is an inevitable property of heat pumps, and especially of vapour-compression heat pumps, that the heat output is affected by the source temperature. This presents the problem that the heat output of an air-source machine decreases as the temperature of the outside air decreases, which is particularly serious for space heating of buildings because the heat demand of the building increases as the outside temperature decreases. Conversely, when the evaporator is unusually warm, the pressure can become high enough to cause the compressor to overload, which is dangerous to both compressor and motor. This is likely to be a particular problem in solar-assisted heat pumps because of the possibility of occasional very sunny days during the heating season. Various means of circumventing this problem have been tried, such as the use of large amounts of boost heating from a source other than the heat pump itself (see Chapter 5). However, such an approach must be regarded as second-best; it is preferable to consider the possibility of modulating the capacity of the heat pump itself.

4.4.1 *Compressor modulation*

One method of modulation is to vary the effective stroke of the compressor, so as to move a greater volume of refrigerant during colder weather when a greater output of heat is required with a lower source temperature. This is fairly easy to arrange with larger compressors, some of which incorporate a variable-stroke facility or offer the possibility of switching one whole cylinder on and off. In smaller, domestic-sized compressors, such refinements are not usually available and it is necessary to consider other methods. Obviously the required modulation can be achieved by varying the speed of the compressor, provided that its efficiency is not adversely affected.

A computer simulation study by Fleming (1978) indicates that it should be possible to match the condenser output to the building heating requirements over a range of outside air temperatures from 10 to 0°C by varying compressor speed by a factor of just over two. Computer studies of compressor valve performance by Bredesen (1979) indicate that there is likely to be some degradation of valve performance at the extremes of a threefold variation of speed (a range of 500 to 1500 rpm was studied) but that it should be possible to obtain a reasonable compromise of valve spring tension which will give good performance for a speed variation of a factor of two. Experimental work by Paul (1979) is in broad agreement with this conclusion. Thus it seems to be possible to contemplate speed variations of the required order without unreasonable degradation of compressor efficiency. Because under mild conditions the heat exchangers are called on to deliver less heat, and thus operate with a lower temperature difference, there is a significant thermodynamic benefit in reduced-speed operation, and the overall coefficient of performance should be significantly improved, especially in climates which feature long periods of relatively mild winter weather.

It is straightforward to operate an engine-driven machine at variable speed, but rather more difficult in the case of electric motors. This topic was discussed in more detail in Chapter 3. In either case it is necessary to derive a signal from a sensor measuring either evaporator pressure or source air temperature, and to apply the resulting control signal to an appropriate motor speed control. A linear relationship between speed and outside air temperature would probably suffice, making the control circuitry simple, though there may be some virtue in applying a more complicated and perhaps empirically derived algorithm.

4.4.2 *Evaporator control*

As an alternative to speed control, simpler methods are available, whose function is primarily to prevent overload of the compressor or motor at high evaporator temperatures, but which achieve some degree of capacity control at the same time. A method used by the authors for air-source machines (McMullan and Morgan 1979) is to sense the condensing temperature and to use it to regulate the effective capacity of the evaporator, by switching or controlling the speed of the fans which blow the source air over the evaporator coil block. A thermistor senses the condensing temperature, and a pair of comparators switches off first one fan, then a second fan, as the temperature rises. Finally if the temperature rises still further, a third comparator applies a fault signal to the protection circuit and stops the compressor. This has been found to work well, especially in air to air machines where the thermal inertia of the condenser is low and it thus warms up quickly, reaching a temperature related to the evaporator temperature and causing the control system to intervene if necessary. In air to water machines, however, it has been found to be less effective, because of the thermal inertia of the water circuit. The heat pump still moves more heat when the source air temperature is high, but the heat demand of the house is less, and the room thermostat is satisfied more quickly before the water circuit has warmed up fully. Consequently the condenser temperature remains relatively low and the control set points are not reached (Morgan and McMullan 1980). It is debatable whether or not it is worth having the control system under such circumstances, although it does give a measure of protection against misuse. A disadvantage of such a control system is that the evaporator is deliberately rendered less effective during warm-weather operation, which reduces the warm-weather COP, but on the other hand it allows the safe use of an oversized evaporator, thus enhancing the cold-weather COP. Overall, the seasonal COP is probably enhanced by the larger evaporator, though at the time of writing the results are somewhat inconclusive on this point.

Another simple method of preventing overloading (McMullan and Morgan 1980) is to measure the current drawn by the compressor motor and use the signal to control the fans in much the same way as has been described above. In cold weather the rate of evaporation of refrigerant is low and the compressor is only lightly loaded, so all the fans are switched on. In warm weather, when the evaporator temperature is high, the motor becomes overloaded, the current becomes too high, and fans are switched off to reduce the effectiveness of the evaporator and thus lessen the load on the motor. A certain amount of capacity control is achieved in addition to the protective function, but the same criticisms apply to this method as to the method described earlier, and its practical advantages must be regarded as similarly suspect.

4.4.3 Condenser control

Calculations by Paul and Steimle (1980) indicate that there can be significant improvements in the actual performance coefficient observed in real instal-lations, as opposed to that measured under constant temperature conditions in the laboratory, if the water flow through a water-cooled condenser can be varied to allow for the greater output of a heat pump when the source temperature is higher. In some respects this is equivalent to the capacity modulation method described in the previous section, and is subject to similar criticisms. However, the calculations indicate a possible improve-ment in COP from about 3.0 to about 4.0 at a source air temperature of 15°C, provided that the water flow can be controlled over a sufficient range, and experiments seem to agree. Unfortunately the range required is very large, about a factor of eleven over the range of outside temperatures from −15 to +10°C, which may present practical difficulties.

4.4.4 Two-motor systems

A very simple control of compressor output can be achieved by having two entirely separate compressor-motor assemblies and switching them on together or separately. There is, of course, a cost penalty, though there is possibly a useful improvement in reliability. The main objection is the likelihood of needing very different sizes of evaporator, condenser, expan-sion valve, etc when operating in the two regimes, and indeed it might be argued that two separate heat pumps would be a simpler solution. If only a single-phase supply of electricity is available, a two-compressor system (or one with two heat pumps) can circumvent the size limitation inherent in single-phase motors by cascaded starting, and this can be an added advantage.

One usually thinks of a two-compressor system working in parallel, both compressors having the same low and high pressures. There can be advan-tages, however, in a series–parallel arrangement. In mild weather, one compressor suffices. In colder weather, two work in parallel. In extreme cold, when the evaporator pressure is very low, the two work in series, thus keeping the compression ratio reasonable. This is an ingenious idea, but is likely to present design problems. It has been argued that such an arrangement would allow heat pumps to be used in colder climates; that may well be so, but the coefficient of performance is likely to be poor.

A more likely two-motor arrangement is the use of a fuel-burning engine for driving the heat pump under extreme winter conditions, when the waste heat from the engine can be used to supplement the output from the condenser, while an electric motor is used for more favourable conditions. This has been considered in more detail in Chapter 3.

4.5 Motor starting controls

In Chapter 3 the desirability of reduced-current starting of electric motors was discussed. Here we consider the means available for switching from 'start' to 'run' conditions.

The most obvious method, applicable especially to single-phase motors, is to adapt the voltage-sensing relay which is already used for disconnecting the start winding. It may be possible in favourable circumstances to use the same relay, not only to disconnect the start winding, but also to short out the series resistor or whatever other method of current limitation is in use. More generally it may be desirable to use separate relays for the two purposes, set to change over at different voltages. In either case, the fact that the changeover is initiated by a factor dependent on motor speed makes for an effective limitation of current transients. However, there is a difficulty that if the supply voltage is too low, or the compressor load is excessive, the speed may never become high enough to induce changeover. At best, this may lead to a burnt-out series resistor, at worst to a burnt-out motor. Clearly a protective system is needed, such as a slow-blow fuse or a bimetal circuit breaker. The authors have used a primitive circuit breaker in the form of a soft-soldered connection to the series resistor—if the resistor gets too hot, the solder melts and the wire flies off. Such subterfuges are doubtless excusable in the laboratory, but are scarcely to be recommended for the production line.

A less obvious method, which is nevertheless common practice for three-phase star–delta starters, is to hold the 'start' condition for a fixed period of time. Three-phase motors can endure the star connection for long periods, so the duration of the 'start' can be several seconds or even longer. Single-phase motors are likely to object to prolonged periods of reduced voltage, especially if the start winding is unable to disconnect because of insufficient speed, so the time period is likely to be reduced to less than a second. It is almost inevitable that on some occasions the start winding will still be connected when the timer switches over, causing a short-term current peak, though a less serious one than that caused by direct on-line starting. On the other hand a burn-out due to slow starting is almost impossible unless the timer fails.

Automatic starting systems are an obvious necessity for internal-combustion engine driven heat pumps. Fortunately they are well developed

because of their use in unattended standby generators. Even more fortunately, they are unlikely to need much attention from the heat pump
designer, who will generally buy in a suitable unit from the engine manufacturer. This subject will not be covered, therefore, in the present book.

No matter what starting system or motive power is used, there is almost
certain to be some restriction on the number of times per hour that the
starter can be used. Single-phase motors tend to have rather resistive start
windings which overheat if used too often, and the associated electrolytic
starting capacitors can also overheat, sometimes with explosive results.
Three-phase motors are more forgiving. The existence of transient current
surges during start-up can cause annoying dips in the mains voltage, and
this is especially a problem in single-phase supplies in rural areas, but it
can be tolerated if the frequency of starts is minimised. Engine-powered
systems take badly to frequent starts, which lead to excessive mechanical
wear of starters and engines, as well as to noise problems. In all cases, the
efficiency of the heat pump is likely to suffer as a result of on–off cycling,
leading to a reduction in the seasonal COP. For all these reasons, the control
system should include protection against excessively frequent starts. Further, the thermostat or equivalent control should be designed to produce
reasonable 'on' and 'off' times in normal operation, and if this cannot be
ensured, there may be a case for the incorporation of a heat storage vessel
such as a hot water tank into the heating system of the building to ensure
a sufficient time constant.

4.6 Thermostats for domestic heating systems

Almost all the room thermostats in use at present consist of a bimetal strip
which flexes with change of temperature, causing an electrical contact to
spring open or closed. The characteristics of such a thermostat can best be
understood by considering its response to a (somewhat unlikely) step
change of temperature. If the thermostat casing is reasonably well venti-
lated, there will be a short dead time due to the thermal impedance of the
casing, during which the bimetal temperature changes relatively little.
Next, the bimetal temperature will rise approximately according to

$$T = T_0 (1 - Z e^{-c/t}),$$

where the Z is the thermal impedance between room air and bimetal, T_0
is air temperature, c is the thermal capacity of the bimetal and associated
items and t is time. This function is sketched in figure 4.9.

It is evident that if the thermostat is to follow the step function closely,
the values of Z and c must be minimised. The thermal capacity c is largely
a function of the mass of metal, and there are practical limits to how far
it can be reduced. The thermal impedance Z can be very considerably
influenced, as can the dead time, by increased ventilation of the case.
Normally this is a matter of allowing enough holes for the air to pass
through, limited mainly by safety requirements, but it is possible to resort

to fanning air through the body of the thermostat, as is sometimes done in greenhouses.

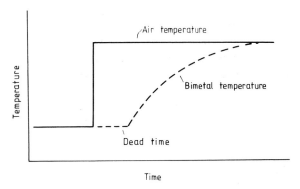

Figure 4.9 Response of a bimetal thermostat element to a step change in temperature.

An alternative method of effectively reducing the time constant is to incorporate an accelerator heater which warms the bimetal strip whenever the thermostat contacts are closed. The dead time and the time constant can thus be drastically reduced. However, the improvement is effective only in preventing overshoot as the temperature is raised by the heating appliance, and is ineffective in the opposite direction. A further disadvantage is that in colder weather, when the appliance is switched on for longer periods, the thermostat is held at a temperature which is significantly higher than the room temperature because of the effect of the accelerator heater, and the result is that the thermostat consistently switches off too soon, and the room is colder than it ought to be. (That does, of course, save energy.)

The other main drawback of a bimetal thermostat is the inevitability of a significant temperature differential between 'on' and 'off'. This is usually about 1 or 2°C, which is small enough to have less effect on precision of control than the time constant of the bimetal. Nevertheless it is a contributing factor to performance limitation. On the other hand, a significant differential provides hysteresis and so acts as an insurance policy against excessively frequent switching on and off; it is by no means a wholly bad feature.

The conventional bimetal thermostat is so familar to heating engineers that it is the obvious choice, albeit a slightly crude one, and its installation is likely to present few problems. However, when used in conjunction with a heat pump there are likely to be a few slight difficulties. First, the accelerator heater is usually selected on the assumption that the circuit to be controlled consumes a certain amount of current, which passes through the bimetal (and in some cases through the accelerator heater) contributing to the selfheating. In a heat pump control system it is likely that the current passing through the thermostat is small, of the order of milliamps if it is switching the coil of a contactor, and microamps if it is connected to an electronic control system. This is quite insufficient to cause any appreciable heating of the bimetal strip, and it is necessary to make up the lost heating effect in some way if the speed of response is to be satisfactory. In the case of parallel accelerators, i.e. heaters which are connected in parallel with the load, it is sometimes possible to remove the existing heater and to replace it with one of lower resistance, or alternatively to add another resistance in parallel with the first and in close proximity to the bimetal. Series heaters, in contrast, require an increase in resistance, and it is sometimes difficult to increase it sufficiently to be effective. Alternatively, it is permissible to put a dummy load across the control system to increase the current through the heater, but the current consumed in this way is sometimes of the order of 0.3 A, which can be considerably greater than the consumption of the rest of the control circuitry, and can necessitate a much larger and more expensive power supply than might otherwise be required.

An electronic thermostat, consisting of little more than a comparator amplifier, a thermistor sensor and a reference potentiometer, can overcome the problems of a bimetal thermostat because of its very small thermal mass, and might even have a somewhat smaller differential between 'off' and 'on', but it is likely to be more expensive than a bimetal unit. It is worth noting, however, that the sensing element in this type of thermostat probably reaches a closer approximation to the actual room temperature than with the bimetal type.

As a more radical departure from conventional thermostat design, it is worth considering the use of a mean radiant temperature measuring system, which attempts to measure the comfort condition in a room, including radiative gains and losses, rather than just the air temperature. Such a system would be especially useful in conjunction with low-temperature radiant heating systems such as an underfloor installation. The traditional means of measuring mean radiant temperature involves the use of a large blackened ball, which is visually unattractive, but it is doubtless possible to devise a miniature version suitable for the living room of a house, though it may be difficult to find an ideal position for it.

Night set-back operation is a very definite advantage with heat pump heating systems. A reduction in the target temperature of perhaps 4 or 5°C is sufficient to reduce the night-time heating demand very appreciably in milder climates, which saves fuel, reduces the running hours and minimises the amount of running at a time when noise might constitute a problem. On the other hand the house is not allowed to cool too drastically, so that a short warming-up time is possible in the morning without needing a considerable and expensive oversizing of the heating system. Night set-back is a standard feature of most bimetal thermostats, often consisting simply of an additional small heater which, when switched on by a timer, raises the temperature of the thermostat slightly above ambient. In an electronic system it would be possible to have two independent reference potentiometers, making possible the precise adjustment of both daytime and night-time temperatures. This might well be no particular improvement over the cruder bimetal version, but it could have greater consumer appeal—especially if the time were controlled by a digital clock!

More sophisticated systems, involving an element of prediction of daytime heat demand on the basis of the temperature during the night before, have certain advantages over simple night set-back systems because the morning warm-up can be brought on earlier in colder weather. Such optimum-start controllers are available, but the voltage of operation may necessitate a relay between the controller and the heat pump.

An interesting problem arises when considering the possible use of local thermostatic controls such as radiator valves in wet systems. In a conventional boiler system the efficiency is not greatly influenced by the area of radiator in use, so it appears to make good sense to use radiator valves as a means of restricting the heat supplied to individual rooms. In the case of a heat pump system, the COP is reduced if the condensing temperature rises, and a restriction of heat dissipation is almost bound to cause such a rise. Consequently it is by no means certain that radiator valves are a good thing to use. Certainly, it is necessary to restrict the heat transfer area controlled by such valves, so that at all times a good area remains effective, and it is certainly not desirable to use such valves instead of a room thermostat directly switching the heat pump on and off. Furthermore, the desirability of local valves has been called into question even in conventional boiler systems in a paper by Rayment and Morgan (1980), who found that a room thermostat switching the boiler directly was no less effective provided the system was reasonably well balanced. Thus the optimum (neglecting cost) may well be to use local valves only as a means of delicate trimming of the balance of the heating in each room, and not allowing them to completely shut off the radiators they control. Similar considerations apply to warm-air systems, though because of the greater

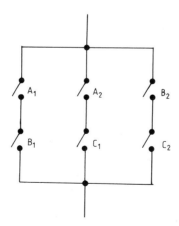

Figure 4.10 Majority-decision thermostat connection, using three two-pole thermostats.

difficulty of ensuring good balance amid the perturbing effects of opening doors, draughts, etc it is probably attractive to compensate for the perturbations by the use of local control in such cases. Furthermore, the major thermal impedance in the 'hot' end of the heat pump is likely to be at the condenser itself in a hot-air system, so it is less likely to be affected by changes in the heat distribution system than in the case of a wet heating system, where the major impedance is at the surface of the radiators in the individual rooms.

If local radiator thermostats are not to be used, there arises a need for a system of overall thermostatic control which is better than a single thermostat but which does not work by restricting the heat dissipation area. One can, of course, conceive of all manner of complicated computer control systems, computing an average value of temperature throughout a building, but an engagingly simple approach to a majority-decision thermostatting system can be wired up as shown in figure 4.10, or alternatively one could consider an analogue averaging connection using temperature-measuring integrated circuits, as shown in figure 4.11. These are likely to give considerably better control than a single room thermostat in a domestic installation, without much extra cost or complication.

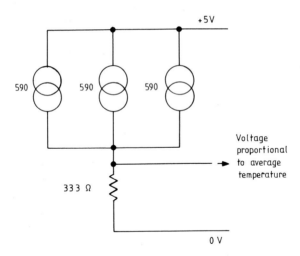

Figure 4.11 Analogue temperature-averaging system, using any number of temperature–current integrated circuits such as the 590 device.

5

APPLICATIONS

5.1 Space heating

5.1.1 *Heating systems and thermal comfort*

The object of any space heating system in a building, whether based on a heat pump or on a more conventional heat source, is to keep the occupants comfortable. Unfortunately the occupants are usually people, rather than thermometers, and this introduces a whole range of variables, many of which are by no means perfectly understood. Much of the research in this field has been directed towards studying the effect of extremes of climate on physiology and on the performance of exacting tasks, usually for military purposes, and is of little direct applicability to more peaceful pursuits.

For non-military personnel an element of individual preference is involved, which must be taken into account despite the difficulty of so doing. A system of comfort voting has been devised by ASHRAE, in which people taking part in an experiment are asked to assess the thermal environment on a seven-point scale: cold, cool, slightly cool, neutral, slightly warm, warm, hot. Numerical values can be assigned to these adjectives on a linear scale, usually 1 to 7 or -3 to $+3$, and an arithmetic mean can be taken from the votes of all the participants to indicate a communal opinion. In order to make observations of this type experimentally convincing, it is customary to set up a very closely controlled environment, to dress the participants in a closely-controlled 'uniform', and to define their activity level precisely. It is also important to define the nature of the participants closely, though this is of course more difficult, and one has to resort to selecting a 'random' group of, say, college-age American males.

Several studies of this nature have been performed, the best known being those by Nevins and colleagues (1964, 1966) for American participants and by Fanger (1970) for Danish participants. The results showed a remarkable degree of consensus; most of the participants agreed fairly predictably about what constituted comfort. Several widely held beliefs about factors which might influence people's opinions were demolished or at least seriously weakened. These included: *nationality*—the hardy Danes and the pampered Americans agreed almost exactly about the ideal temperature; *age*—elderly and college-age people had similar preferences, though that does not mean, of course, that elderly people have as good a tolerance of extremes; *sex*—men and women clad in similar clothing agreed about the optimum, though women were more sensitive to non-ideal temperatures; and *body build*—fat and slim people, while seated, agreed about the optimum.

From the viewpoint of the heating engineer, this is of course very useful, but it is of little specific relevance to heat pumps. Of more use for the present purposes are studies which have been made of the response of participants to different methods of applying heat to a heated space: low-temperature radiation, high-temperature radiation, convection, etc; and also to non-uniform heating and cooling: warm floors, warm ceilings, draughts, etc. Although experiments with real people are notoriously unsatisfactory, it seems that people in general feel they are able to tolerate a lower air temperature if they are bathed in low-temperature radiant heat from an extended surface such as a floor or ceiling, and that they can often be more comfortable under these conditions (ASHRAE 1973). A warm floor seems to be generally acceptable provided it is not hotter than about 30°C, and it also seems to be more acceptable than a warm ceiling (Fanger 1970), an effect which may be attributable to the levelling out of convective temperature gradients. For the heat pump engineer this is of considerable interest, since it should be rather easier to heat an extended surface to say 30°C, implying a condensing temperature of perhaps 35°C, than to heat a smaller surface such as a metal radiator to 60°C implying a condensing temperature of perhaps 65°C, and it might also be rather easier to pipe warm water into a floor than over a ceiling. In the past there has been considerable doubt about the ability to transfer enough heat from a floor or ceiling heating system to satisfy the heat demand under the worst conditions, but in the wake of the energy crisis it is now customary to require much better thermal insulation in the building envelope, and the required heat transfer can usually be achieved without difficulty in most up-to-date buildings. This can be shown by some approximate calculations.

Convective heat transfer from a heated floor, in the absence of forced draught, is given by

$$q = K(t_f - t_a)^{1.31},$$

where q is heat transfer in W, K is a constant, approximately 1.82 W/°C/m², t_f is floor temperature in °C and t_a is air temperature in °C (modified from ASHRAE 1973). Thus at an air temperature of 20°C and a floor temperature of 30°C, the heat transfer rate by convection is 37 W m⁻². Radiative heat transfer is given by Stefan's law:

$$q = \sigma\varepsilon(T_f^4 - T_a^4),$$

where q is as above, σ is Stefan's constant $= 5.67 \times 10^{-6}$ W K⁻⁴ m⁻², ε is the emissivity (lying between 0 and 1) and T_f and T_a are as above, but in *absolute* temperature (kelvin). For a floor of emissivity 0.5 at a temperature of 303 K (30 °C) and for a mean radiant temperature of the surroundings of 293 K (20 °C) this gives a heat transfer rate by radiation of 30 W m^{-2}.

Both of these figures are rather approximate, and the reader is referred to the detailed tables published by CIBS (1980) for more precise data. Nevertheless, the agreement with the tabulated data is good. Thus the heat output is a useful balance of radiation and convection, and totals roughly 70 W m^{-2} which is adequate for buildings built to more recent standards of insulation.

As a consequence, there has been a rapid development of floor-heating systems, notably in Germany, using plastic pipes buried in a special concrete screed (Field 1976). Some disadvantages may exist, for example there is usually an element of thermal storage in any floor heating slab, which may render air temperature control less precise under conditions of fluctuating outside temperature, and this can be a serious problem in lightweight buildings with solid concrete floors, but for most normal building envelopes the results should be acceptable and construction techniques have been introduced which minimise the amount of storage, thus rendering the problem minimal. Consequently one can be fairly definite in saying that such systems are ideal for heat pumps.

In an existing building, it is not usually possible to fit underfloor heating without unacceptable disruption, so the engineer's choice is more restricted. Hydronic systems using the traditional pipes and radiators are quite easy to install, but it is by no means clear that they will work well at the low water temperatures favoured by the heat pump. The radiative heat transfer, being proportional to T^4, will fall drastically with a reduced water temperature, while the convective heat transfer, which (for a vertical panel) is proportional to $\Delta T^{1.33}$ (CIBS 1980), is less sensitive to water temperature but is still reduced considerably. It is rather difficult to calculate the exact effect from general equations such as Stefan's law and the convective heat transfer equation, because the proportion of radiative and convective heat transfer depends very considerably on the design of the radiator surfaces. Instead, it is preferable to consult manufacturers' charts of heat output for a given design. For a typical panel radiator (Veha, made by Associated Heating Engineering and Ventilation Ltd) the tables indicate that if one reduces the temperature difference from the 'normal' value of 60 °C to a lower value of 30 °C, the heat transfer is reduced to 0.4 of its 'normal' value. Thus if one compares radiators at 80 °C in a room at 20 °C with the same radiators at 50 °C under the same conditions, the heat emission is reduced to about 40%, implying that the radiator size in a low-temperature system must be increased to about 250% of its 'usual' area. This is enough to horrify most heating engineers and to antagonise the occupants of the building. This is not the whole story, however, because the nature of the duty cycle of the heat source must also be considered. Boilers are cheap, and it is customary to install quite large boilers in quite small buildings, to ensure that there will always be enough heat. As an example, a house of 100 m², which in typical cold winter weather might require 5–7 kW of heating, might easily be fitted with a boiler of 18 kW (60 000 Btu/h) output. This means that enough radiator area has to be provided to emit 18 kW at reasonable water temperatures. Heat pumps, in contrast, are expensive and it would be uneconomic to buy one whose output was significantly larger than the maximum demand expected. Consequently it is necessary to install radiators with enough area to dissipate only about 7 kW at 50–60°C, which by a happy chance is almost exactly 40% of 18 kW. It appears, therefore, that a radiator installation suited to a typical domestic boiler should be well suited to a typical domestic heat pump. This sounds too good to be true, and there is, of course, a catch—the boiler and its radiators have enough spare capacity to warm the house up from cold even in the worst weather, whereas the heat pump has little in reserve for such conditions. It is preferable, therefore, to use night set-back controls (Chapter 4) with heat pumps rather than to turn the machine off completely on cold winter nights.

The remaining alternative, an air distribution system, is a favoured option in America where the need for air conditioning in summer has tended to encourage the use of air distribution both for cooling and for heating. However, it is probably not ideal. For maximising the COP it is necessary, of course, to use a low air temperature, which implies a need to move rather large air volumes. This is a certain recipe for complaints about draughts. Nevertheless, with careful installation, it can undoubtedly be used successfully, especially in well insulated buildings where the amount of heat needed is small. Various precautions against the spread of smells and the carriage of sounds are needed, and the reader is referred to sources such as ASHRAE for details. Another problem which is attracting attention is the fire hazard associated with blown-air systems. It is possible for the heating system to feed fresh air to a fire for some time unless fairly extensive—and expensive—measures are taken over fire detection and the inclusion of fire dampers in the duct work.

5.1.2 Supplementary heating

As an example of other problems encountered in space heating applications, consider domestic heating. Here, the size of the heat pump has to be optimised to minimise capital costs while maximising the utilisation (and usefulness) of the installation. Several psychological or social factors are more important than technical ones. For example logic tells us that the heat pump should be sized to meet all of the heat demand only down to some selected design temperature. (This is distinct from the previous point about the excess capacity usually included in domestic systems to provide fast warm-up from cold). On the very few days in the year when the temperature falls below this value, the deficit is made up by a supplementary heating supply. In the UK a suitable choice would be between 0 and 2°C as, over most of the country, the mean temperature falls to less than zero on only nine days in the year, while it is less than −3°C on only about one day per year. Unfortunately, however, customer reaction to being provided with a heating system which needs to be supplemented when it gets cold can be expected to be poor. The customer has purchased a *heating system* and will be disappointed if it does not meet his heating demand. He will be reluctant to add to his system unless very strong arguments can be given.

This objection is conventionally met by the provision of boost heaters which are switched on to provide the necessary back up when required.

Part of the application design philosophy lies in deciding the balance temperature of the heat pump system, the amount of boost heating to install, the switching mechanism (manual or automatic, and on what criteria), and whether or not a measure of thermal storage should be included.

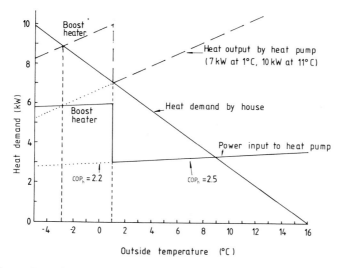

Figure 5.1 Comparison of heat demand of house with supply from heat pump and 3 kW supplementary heater.

Since most domestic heat pump installations will probably use air as the heat source (and since, in any event, the house is exposed to the air), the main difficulty governing the performance of heat pump units will be the extreme variability of air temperature, even on a short-term basis. This means that the performances of the house and of the heat pump are constantly changing (and in opposite directions). Organisations such as CIBS in the UK and ASHRAE in the US publish data whereby it is possible to make statistical estimations of the behaviour of the house after taking into account the incidental gains arising from solar radiation, occupancy, cooking, lighting, etc, and a graph of probable heat demand against air temperature, wind speed and so forth can be plotted. On this for comparison can be superimposed the heat output of the suggested heat pump as shown in figure 5.1. Here, the house is assumed to have a design heat load of 8 kW at −1°C, and to be equipped with a heat pump which has an output of 7 kW at +1°C. In order to parametrise the model a little better, the heat pump is assumed to have a COP_h of 2.5 at 7°C and 2.2 at 0°C, and 3 kW of boost heating is included. It is apparent that the heat pump alone will meet the total heat demand down to the balance temperature of 1°C. Below 1°C the boost heater will be needed, and will meet the demand until the air temperature falls below −3°C. At this point further supplementary heat is necessary.

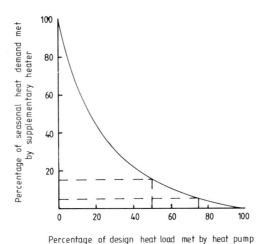

Figure 5.2 Comparison of supplementary heating with portion of design heat load met by main heating system.

It is clear from this discussion, that in every case a decision must be made regarding the balance point of the system and how much boost heating to provide. Here, weather statistics are essential, and once again the CIBS and ASHRAE data provide the necessary information. In particular, it is possible to estimate the amount of supplementary heating that will be required over the season as a function of the fraction of the design heat load that is met by the heat pump (see figure 5.2). Thus, if the heat pump is sized to meet 50% of the design heat load (−1°C) then 15% of the total heat demand will need to be supplied by the supplementary system. If the size of the heat pump is increased to meet 75%, however, then the supplementary heat requirement falls to only 5% of total heat demand. It is apparent that the matching of the house and system is fairly critical. The example of figure 5.1 is continued in figure 5.3, where the influence of the normal temperature distribution on the seasonal energy demand and heat supply of the system is shown. The relative unimportance of the period when temperatures fall below −3°C (on average) is readily apparent, as is the influence of the supplementary heater on the total power consumption of the system.

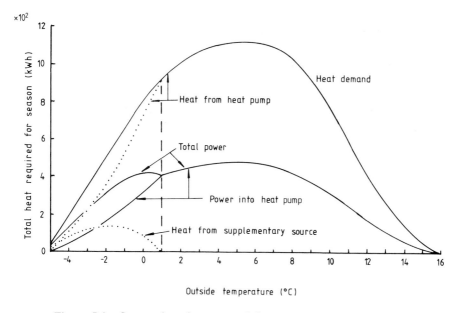

Figure 5.3 Seasonal performance of the system shown in figure 5.1.

The same difficulties arise with any heating system, but with oil- and gas-fired boilers it is easier and cheaper to provide the extra capacity—essentially by changing the burner nozzles. This flexibility is not open to the heat pump designer because of the much greater cost of changing components and, as a consequence, system design takes on a significance previously not often encountered in the domestic market.

5.1.3 *The effect of storage*
Heat pumps are usually electrically powered, and the electricity industry has become so attached to encouraging the sale of off-peak power that the question invariably asked is, can one devise an off-peak heat pump system with suitable storage? Obviously one can, but it may not be sensible to do so. Most off-peak tariffs offer an eight-hour period of availability at most, and if this is to provide the same amount of heat as an unrestricted (24-hour) tariff, one can expect to need a heat pump of about three times the heat output. Furthermore, the off-peak tariff is available only at night, which for an air-source heat pump is undesirable because that is when the air temperature is at its lowest, thus giving the worst COP. Thus it would require an immense difference between the prices of electricity on the two tariffs to make an off-peak heat pump economic, at least for ordinary space-heating purposes. The only exception to this is where the heating requirement is itself a nocturnal creature, as might be found in horticultural glasshouses.

Nevertheless, the use of off-peak heat storage as a component of a heat pump system should not be completely rejected. It is quite conceivable to use small amounts of stored heat as a boost for coping with unusually heavy heating loads, such as the morning warm-up, which, since it occurs immediately at the end of a night-time tariff period, can use stored heat conveniently and efficiently. (Larger quantities of storage would be needed to cope with cold snaps in winter.) In another traditional off-peak market, that of heating domestic hot water, there is the possibility of using the heat pump to preheat the cold water prior to entering the hot-water storage cylinder, thus saving some of the off-peak power but not eliminating it. In this case it is important from the economic point of view to ensure that the COP of the heat pump is high enough to reduce the effective cost of the heat to a figure lower than that of the off-peak power which it replaces, otherwise no saving of money will be possible.

Consequently it is unwise for electricity supply authorities to view heat pumps as a threat to their off-peak business. Instead, they should look for ways in which they can sell off-peak power as an adjunct to a heat pump, providing the prospective customer with an attractive package deal and providing themselves with a reasonable load factor.

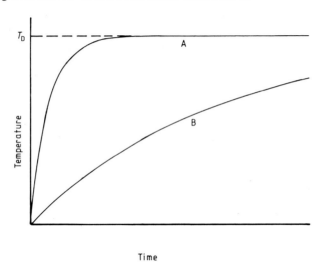

Figure 5.4 Effect of time constant on radiator temperature. A, short time constant; B, long time constant. T_D is the design radiator temperature.

Off-peak storage is not the only way in which an element of storage can benefit a heat pump installation. Consider for example the case of a hydronic heating system with a fixed-speed heat pump incapable of effective capacity modulation. When the building calls for heat, the machine switches on, warming the radiators and eventually satisfying the room thermostat, whereupon the heat pump switches off and the radiators cool. If the thermal capacity of the hydronic system is small, the radiators warm up quickly to their normal operating temperature and thus they spend a long time at a

112

high temperature when the COP of the heat pump may be rather poor. In contrast, if the time taken to warm up the radiators is long compared with the time constants of the building and the thermostat, the radiators tend to spend most of their time at an intermediate temperature (figure 5.4) when a good COP can be expected. It is conceivable that a significant improvement in seasonal COP could be achieved by deliberately including a small amount of thermal storage to increase the time constant of the radiators in a hydronic system. Unfortunately it is not easy to predict the effect of such a modification, and work is in hand in the authors' laboratory to study the problem by computer modelling. Nevertheless, it looks promising.

5.1.4 *Component selection*

Let us imagine that the reader has been asked to design a heat pump installation from the beginning, and follow the problem through.

Obviously one must begin with the heat load—how much heat is needed, and how the demand varies with time of day, weather, etc. At the same time it is worth looking round to see if a useful cooling load exists in the vicinity, and to estimate how much cooling is needed and how that load varies also. Unfortunately it is unusual for either heating or cooling loads to be known in detail, and it is necessary to assess them with the aid of the publications of ASHRAE, CIBS etc and with the assistance of a friendly building services engineer.

Next, one should probably look at the heat transfer system, to establish what the customer needs and can afford. This involves a selection of the fluid to be moved (air, water, brine, or perhaps even Freon) and the acceptable range of temperatures. At this stage also one should consider the heat source, unless of course one has already earmarked a cooling load for the purpose. Again, this involves a choice of fluid (air, ground water, river water, . . .) and a consideration of the likely range of temperatures. It hardly needs to be restated that one should normally look for the warmest source that is available and for the coolest heat transfer system that is acceptable.

One is now in a position to choose a refrigerant. Usually there are several which have suitable ranges of temperatures and pressures, and the choice may well not be very critical, but it is worth spending a little time with the thermodynamic charts to ensure that nothing unexpected happens over the expected ranges of temperatures. One should try to avoid wet compression, exceeding the critical temperature, excessively high pressures, pressures below atmospheric (unless one is confident of one's soldering), and excessive superheat temperatures at the exit from the compressor. If these needs are satisfied, it is worth calculating the COP at a few pairs of reasonable evaporating and condensing temperatures, for several refrigerants, assuming that all the operations are thermodynamically ideal. This may lead to a preference for one refrigerant rather than another. Finally one should check that the refrigerant is actually available (they usually are, though some are more readily available than others), how much it is likely to cost, and whether it is safe.

One can now choose the compressor. Most refrigeration manufacturers offer catalogues which contain graphs or tables of heat extraction (the cooling effect) and motor power consumption against evaporating temperatures and condensing temperatures, for several of the common refrigerants. These are usually for one particular operating condition, such as a fixed suction gas temperature of 18°C and no liquid subcooling, and it is unwise to assume that these conditions will prevail in one's own design, but nevertheless the compressor figures will usually allow one to select a machine with reasonable accuracy. Since it is usually the heat extraction which is tabulated, rather than the heat rejection, it is necessary to add on the power put into the system by the compressor and motor. This will not necessarily be equal to the power consumption, since there will be some losses, but for the present purpose it will usually suffice to take the losses as zero.

Next comes the motive power. For most electric compressors this has already been chosen by the makers, and it is sufficient merely to check that the motor is unlikely to suffer overload at any likely combination of temperatures. For open compressors and non-electric systems, there is much more flexibility, but one should ensure that enough power is available under all likely conditions and, equally important, that there is enough torque at all likely speeds of rotation. For the latter one needs not only a torque–speed curve for the motor, but also one for the compressor, and it is unfortunately unlikely that both are available. At this point the foolhardy will opt for direct drive, the faint hearted will select belts and pulleys so that the gear ratio can be changed, and the conscientious will initiate a series of experiments to find the best ratio. (The authors are probably in the first group!)

One can now select the expansion valve. If a common refrigerant has been chosen it should be easy to find a valve with a suitable bulb charge, and one simply chooses a valve and an orifice size to match the heat extraction. Usually this is not too critical because there is a substantial control range, but if a valve proves to be unsatisfactory in service, one can change the orifice easily and cheaply. If an uncommon refrigerant is in use, it will be necessary to ask a valve manufacturer to make a valve with a suitable charge, and they are usually quite willing to do so, or one can change the charge oneself—it is not particularly difficult. Alternatively, another type of control such as a float valve or an electric valve may be a more economical solution.

The last major components are the heat exchangers. These are described quite adequately in manufacturers' catalogues, and can be selected on the basis of the required heat extraction and rejection, bearing in mind the temperature differences between the refrigerant and source fluid or the heat transfer fluid. It is worth pointing out that from the scientist's viewpoint there is no such thing as (say) a 10 kW heat exchanger, since the heat transfer depends on the temperature difference between the fluids, whereas manufacturers tend to catalogue their products simply in terms of their heat transfer capacity. Obviously, therefore, it is important to ensure that the temperature difference being used as a standard for comparison is actually a reasonable one. In the authors' experience, it is often wise to err on the generous side in provision of heat exchangers, especially in the evaporator of an air-source machine. One common query which arises in heat exchanger design is whether or not a countercurrent flow is desirable. Because the evaporation and condensation processes are (in principle) isothermal unless one uses the Lorenz cycle (see Chapter 2), there is not much benefit to be derived from a countercurrent flow in those parts of the cycle. However, the superheated and subcooled regions of the cycle are not isothermal, and in those regions there can be advantages in countercurrent flow. In particular, the desuperheating of the vapour from the compressor, prior to the condensation process, can yield dividends in the shape of a higher temperature for the heat supplied to the load. Usually this is enough of an advantage in itself, but if it also allows the condensing temperature to be reduced, an increase in COP can be achieved. Thus it is definitely advantageous to use a countercurrent condenser, or alternatively, to use a countercurrent heat exchanger for desuperheating the vapour prior to its entry into a non-countercurrent condenser. At the other end of the cycle it is less important to use a countercurrent evaporator, unless one is particularly anxious to lower the temperature of the source fluid as far as possible. In general it is comparatively easy to arrange countercurrent flow when the fluid other than refrigerant is a liquid. It is considerably more difficult to arrange when the other fluid is a gas, though a degree of countercurrent is possible in a deep coil block by feeding the refrigerant in at the face of the block from which the other fluid is exhausted, and taking the refrigerant away at the other face. It might appear from this that a deep coil block is desirable, but as it is more difficult to defrost a deep block, the advantage is dubious. The only remaining place in which a countercurrent heat exchanger might be useful is in subcooling the liquid coming out of the condenser. This is often pleasingly warm, and it is annoying to be unable to use the heat in it. Unfortunately it is usually at too low a temperature to be of use.

5.2 Heat Recovery

The preceding discussion on space heating assumes that the heat source is one of the conventional naturally occurring sources. In larger scale applications, such as commercial premises, swimming pools, etc and in many industrial cases, however, considerable heat is being wasted through the necessity for ventilation, or through the disposal of waste process heat. One approach to conserving at least part of this is to install heat recuperation equipment which transfers heat from the discharge stream to the inlet stream. Where applicable, this is extremely effective and energy savings of between 50 and 80% are achievable. The disadvantage of conventional equipment, unfortunately, is that it must act essentially as a feed heater, and the recovered heat is of a lower quality than that being wasted. The heat pump provides a route towards achieving the same degree of recuperation or greater, with the added advantage that the heat can be returned to the system at a more directly useable temperature. In addition, it is possible to design a heat pump system to optimise the recovery of latent heat as sensible heat, which is not always possible with direct heat exchanger systems. For example, in this context, the thermal wheel may well condense water vapour out of the exit air stream from a building, but it then returns it as water vapour to the inlet stream. That is, the latent heat has been returned as latent heat. Runaround coils will transfer latent heat to sensible heat to a certain extent, but only the heat pump can be arranged so as to actively condense water vapour on the evaporator while transferring all the heat so collected to the condenser as sensible heat. Other possibilities occur in any commercial or industrial application where there is either an established cooling requirement, or an unacceptable heat load in one part of a process.

The classic example of heat recovery of this type probably lies in the opportunities offered by covered municipal swimming pools. Here many of the essential ingredients are combined: a large and fairly stable heat demand, a requirement for mechanical ventilation both for comfort reasons and to keep air-moisture levels down to protect the fabric of the building, and a centralised installation for the plant. Over the last number of years, the criteria to be met in designing municipal swimming pools have changed markedly, with a consequent increase in energy demand. For example, older pools were designed for air and water temperatures of 21°C, while the present standard is for a water temperature of 28°C and an air temperature of 27°C. When this is coupled with the increased ventilation rate necessary to maintain spectator comfort and to protect the building (about 10 air changes per hour or about $0.015 \, \mathrm{m^3 \, s^{-1}}$ ($\mathrm{m^2}$ of wetted area)$^{-1}$, the evaporation rate from the pool is about doubled compared with an older pool—to about $2400 \, \mathrm{l \, m^{-2} \, yr^{-1}}$. This corresponds to a latent heat loss of about $1650 \, \mathrm{kWh \, m^{-2} \, yr^{-1}}$. Thus, a municipal pool may suffer a heat loss of about $650\,000 \, \mathrm{kWh \, yr^{-1}}$, which is needed purely to evaporate water from the wetted surface areas. If a sizeable proportion of this (say 75%) can be recouped together with about 50% of the sensible heat input to the pool hall air, then savings of up to two million $\mathrm{kWh \, yr^{-1}}$ are possible. An example of such a system is shown in figure 5.5.

Other examples of heat recovery of this type include the ice rinks used for the 8th Winter Olympics in California, where the heat extracted from the ice rink was used to heat the buildings, preheat the hot-water supply to 38°C and to clear snow off the roof.

Very similar problems arise in department stores and supermarkets where heat is required, where there may be a simultaneous cooling load for freezer cabinets and cold stores, and where the customers provide an additional and variable heat and moisture burden which is usually handled by ventilation. Heat pumps provide a way of reducing the heating bill and of improving the performance of the building.

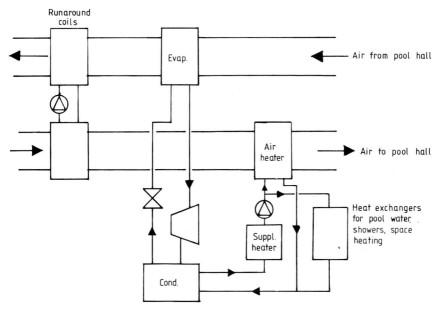

Figure 5.5 Swimming pool heat recovery system.

One particularly elegant use of heat pumps in this way for large office buildings is the Versatemp system developed by Temperature Ltd. In its simplest form this consists of a number of individual room heat pump/air conditioner units, each with its own controls, yet each linked to a common water distribution system which is maintained at 27°C by a centralised boiler. If one individual room is too warm, the heat pump switches from heating to cooling mode and rejects the heat collected from the room air to the water system where it raises the temperature of the return water supply to the boiler. Ultimately, if one side of the building is overheated while the other is too cold, which for example could happen because of solar gains, then heat will be effectively moved from the hot side to the cold side, and there will be a reduced, or even zero, load on the boiler. Obviously, the heating boiler must be capable of supplying all the heat pump units under extreme heating duty, and conversely, there should be a facility for sufficient heat rejection to supply full air conditioning in the summer, if desired.

In industrial heat pump applications, while there is a need for heat recovery from space heating exactly as with other fields, the distinguishing area is the recovery of waste process heat. Here, the range of applications is very large; so large, in fact, as to defy classification in a short space and

117

the interested reader is referred to Reay (1979) for an extensive discussion of the field. In summary, however, the waste heat can be at a low or high temperature, can be air or water, or even warm manufactured products that must be cooled, and can be available either on a batch process or a continuous basis. Nonetheless, a small number of common threads are identifiable.

The standard process heat distribution medium is steam. The reasons for this are obvious and revolve around the central fact that 1 kg steam carries with it the latent heat of vaporisation (2250 kJ), while 1 kg water at the same temperature does not. For this reason, while there is a considerable interest in recovering any waste heat, there is an even greater interest if the heat can be recovered in such a way as to produce low-pressure steam directly. This is an area of primary interest for heat pump development and is one that will reap rich rewards. At the moment, the best known commercially available heat pump which operates at temperatures into the low-pressure steam region is the Westinghouse Templifier. This unit has fairly large heating capacities of up to 3 MW, and is based on a centrifugal compressor. The manufacturers claim that it will produce water at 100 °C from source water at 78 °C with a COP_h of 6.0. The smallest unit in the range has a rated motor power of 63 kW and the largest is rated at 612 kW (source inlet 35 °C, outlet 30 °C; delivery inlet 54 °C, outlet 66 °C). The respective heating duties are 265 kW and 3 MW.

Despite the importance of steam production, there are a number of interesting potential applications at much lower temperatures, and some of these arise because there is a current legal or other requirement for cooling. For example, in a number of areas it is compulsory for factories to cool effluent to some 'environmentally acceptable' level before rejecting it to a river. Presently, this is achieved by either water–water or water–air heat exchangers, and represents a required cooling cost. It also raises the curiosity that the total environmental heat load is unaffected by the cooling so that in absolute terms nothing has been achieved. We have therefore one of the ideal conditions for applying a heat pump—an existing requirement for cooling. In this sort of case, the cooling could possibly be achieved even more effectively by a heat pump, and useful heat could be provided for heating offices etc. Many industries involve variations on this theme: refrigeration/washing, air conditioning/space heating, motor cooling/boiler feed heating, and opportunities can be readily identified in most industrial operations.

A good example is provided by a recent study of the possible energy saving to be achieved on farms by using the heat from milk chilling operations to heat water for washing (Ubbels *et al* 1980). The study was made on farms of up to 85 cows and the milk chiller involved a heat pump and a milk precooler which transferred heat from the warm milk (34 °C) to the supply water (11 °C), bringing the milk to 21 °C and the water to 20 °C. The heat pump then chilled the milk and supplied hot water at temperatures up to 80 °C. The study indicated that in the Netherlands, with milk production at about 11 Mtonne/yr and with an electrical requirement of 500 GWh yr^{-1}, of which one third is needed for cooling and two thirds for heating water, a saving of 200 GWh yr^{-1} (40%) was possible by using this heat pump system.

5.3 Moisture Removal

The removal of moisture is important in a number of application areas, as has already been implied several times. There are two distinctly different types of moisture removal. One, in which the problem is to eliminate air-borne water vapour (department stores, swimming pools, etc), and another in which liquid water is to be extracted from some material. In fact, it is not necessary for the substance being removed to be water, but this is the most common application.

The differences between the two cases are obvious. In the first, the heat has already been supplied from somewhere to evaporate the water and increase the air humidity; in the second, the water is still present as liquid, and part of the drying process is to supply the heat to vaporise it. Both are amenable to heat recovery techniques, and the first has already been outlined for swimming pools in the previous section.

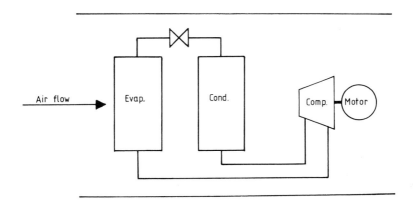

Figure 5.6 Dehumidification using a heat pump.

Air dehumidification can be achieved by three processes: sorption, air compression and refrigeration. In the sorption process, a dessicant material is used to soak up the moisture and subsequently the dessicant has to be heated to drive off the water it has collected. Air compression allows the reduction of absolute humidity if the air is compressed, cooled back to ambient temperature, expanded back to ambient pressure and then heated to ambient temperature again. Because of inefficiencies in the expander, this process is not very effective. Refrigeration allows the air to be cooled below the dew point so that water vapour condenses and runs off. Subsequently, the air has to be reheated to the desired ambient temperature.

119

The heat pump allows the recovery of the latent heat of evaporation as shown in figure 5.6, by first passing the moist air from the room over the cold evaporator coil, condensing out the water and reducing the absolute humidity. The air is then passed over the hot condenser coil where both the sensible heat removed to lower the temperature below the dew point and the latent heat of condensation of the water vapour are added to it as sensible heat, together with the work of compression. Finally, the air is passed over the compressor itself, to pick up any heat losses. Typically, a dehumidifier of this type will have an *effectiveness* of about $1 \, \text{kg} \, H_2O \, \text{kWh}^{-1}$ at an ambient temperature of about 20°C, and relative humidity of 80%. That is, it will remove about 1 kg water from the air for each kWh of electricity supplied to the compressor. In the UK about 7 kg of moisture are generated within a house each day.

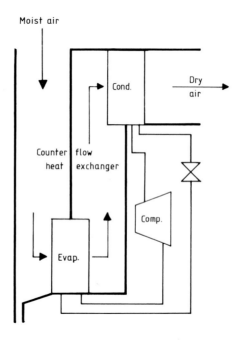

Figure 5.7 Geared heat pump dehumidifier.

An interesting variation is what Blundell (1978) has called the *geared heat pump*. Here, as shown in figure 5.7, the air flows to and from the evaporator are passed through a counterflow heat exchanger. This provides further cooling of the inlet air and consequently extra dehumidification compared to using the evaporator alone. More fan power is needed to offset the extra impedance, but the overall effect is to triple the moisture removal capacity, and double the effectiveness.

The simplest way of drying materials such as timber, paper etc, is to pass heated air over the moist material and then vent the moist air to the atmosphere. Sometimes some of the vented air is recirculated to the inlet, with some saving in energy cost, but with an increase in drying time. In crop drying, it is customary in the UK merely to fan unheated air through the material and dry to whatever equilibrium is attainable under the conditions at the time. If it is a wet season, then obviously the air humidity will be high and the equilibrium crop moisture levels correspondingly high. Since about 2.5 MJ kg^{-1} has to be supplied to evaporate water, and since all of this moist air is vented, there is obviously a potential opening for heat recovery.

As with air dehumidification, the heat pump offers a clear advantage over other approaches. Warm dry air is blown through a kiln containing the material to be dried, picking up moisture and becoming warm moist air. This can then be passed over the evaporator of a heat pump, cooling below the dew point exactly as before (figure 5.6) and condensing out the water vapour which then runs off. The dried air now passes over the condenser and compressor, recovering the latent and sensible heats, and adding the compressor power, which becomes the only energy input to the system. The warm dry air is then passed over the material to be dried . . . Effectivenesses of about 3 kgH$_2$O kWh^{-1} have been reported, which compare very favourably with the 0.3–1.0 achieved by traditional methods. Currently, kiln temperatures of about 50 to 80°C are used, but there is considerable pressure to increase the temperature to 100°C and above.

One of the areas which has attracted most attention for the application of heat pumps has been the drying of timber. In France, there are about 1000 heat pump timber driers and the effectiveness is reported as being about 2 kg kWh^{-1}. Probably more important than the actual energy saving for these applications involving high-quality timber is the control of conditions that is afforded by the heat pump, though German and Belgian studies claim energy reductions by factors of between three and five, and comparable drying times, when compared with traditional high-temperature driers.

At the other extreme, in the drying of crops, it is likely that, since the tolerable temperature rise is small, recirculation might not be worthwhile, so that a very simple open cycle drier is attractive. In this case, the evaporator can be placed in the warm damp exhaust air stream, and the condenser located in the inlet air stream. Thus, the latent heat of the extracted moisture is transferred to heat the incoming air stream, reducing its relative humidity and so increasing its moisture removal capacity. Experiments with this sort of system in the authors' laboratory have shown effectivenesses of about 2.2 kg kWh^{-1}.

The third area is the distillation or liquid concentration problem. This can be tackled in two ways. One is to use a heat pump which heats the liquor, evaporating some water which is then condensed on the evaporator surface as with all of the systems above. This is then transferred by the heat pump to the condenser to maintain the elevated temperature of the liquor as shown in figure 5.8.

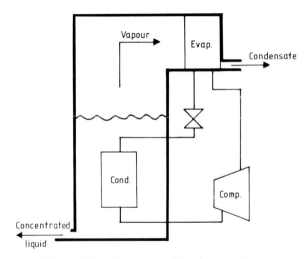

Figure 5.8 Heat pump liquid concentrator.

The alternative is the long established practice of mechanical vapour recompression which deserves greater recognition because of its proven high efficiency. Water extraction rates of up to $22 \, kg \, kWh^{-1}$ have been shown to be feasible.

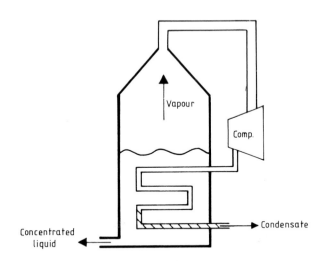

Figure 5.9 Mechanical vapour recompression.

The system is as shown in figure 5.9. The liquid is heated, becoming more concentrated as the water (or other) vapour is driven off. The vapour, meanwhile, is removed by a compressor, which compresses it and raises its temperature. The hot vapour is now condensed in a heat exchanger immersed in the mother liquor, releasing latent heat to maintain the elevated temperature. In practice, only fairly low temperature differences are required and very high COPs (of up to 10) are to be expected. Because the mode of operation favours the displacement of large vapour volumes through small temperature differences, centrifugal or turbine compressors are likely to be preferred to reciprocating units, which raises the problem of matching compressor specification to the duty.

5.4 Economic assessment

The only reason why there is a rising interest in heat pumps is that there is an economic and social pressure to reduce fuel consumption. This is true in all countries, and in every sector of the economy. As a consequence, it is important to be able to assess the likely success that heat pumps will have in contributing towards this goal. Success can be measured in two ways: in terms of reduced energy consumption, or in terms of reduced expenditure. In principle, either is acceptable, but unless both are demonstrably achieved, the heat pump will be in the same position as a number of other superficially attractive projects and will be doomed to failure—and deservedly so. Let us briefly discuss both methods of assessment.

5.4.1 *Economic analysis*
In industrial practice a large number of elements are considered when determining the economic viability of any project. These include the size of the investment, the fuel cost saving each year, the lifetime of the plant, the annual maintenance costs, the rate of inflation, operation costs, etc. There is considerable divergence of opinion regarding the factors to be included in any such analysis, but for R and D and R, D and D projects three approaches are most commonly used: (i) the payback period (PBP), (ii) the rate of return (ROR), and (iii) the present worth (PWV). None of these is terribly sophisticated. All three are open to criticism and are subject to considerable uncertainty, but each in its own way gives at worst a first approximation to the correct answer.

The *payback period* is the simplest and is used extensively. It is defined as

$$\text{PBP} = \frac{\text{initial investment}}{\text{net annual savings} - \text{annual depreciation}}.$$

The initial investment is the full cost of the new installation including the costs of planning etc. The net annual saving is the difference between the energy cost saving and the annual cost of owning and operating the plant. These costs include interest charges, maintenance costs, insurance, etc. Annual depreciation can be most simply estimated by dividing the capital cost by the expected lifetime of the equipment, and inflation can be included if desired. Frequently, a much simpler equation is used. Costs and depreciation are ignored so that PBP = initial investment/annual savings.

Obviously this gives an artificially low estimate of the payback period, but it can be of use when comparing two systems when each incurs maintenance and other costs, but one costs more than the other yet is less expensive to operate. In this case, the payback period associated with the energy saving is defined as ΔI/annual saving, where ΔI is the difference in capital cost between the two plants.

The *rate of return* on investment is basically defined as the inverse of the payback period and is given by

$$\text{ROR} = \frac{\text{savings} - \text{depreciation}}{\text{investment}}.$$

This is frequently calculated only for the first year of operation to give an indication of performance. A variation known as the internal rate of return is often used, which has its origin in discounted cash flow accounting, and which is the rate of return such that, after the lifetime of the installation, its worth will be zero. Internal rate of return, r, is defined by the equation

$$\Delta I = \sum_{n=1}^{N} \frac{\text{saving in year } n}{(1 + r)^n},$$

and has to be calculated by an iterative method.

The *present worth* approach is based on an attempt to estimate how much one should be willing to invest now to achieve savings in the future. This is done by making an estimate of the total value today of the future savings, the *present worth*. If the capital cost of a new project is less than this estimate, then the project is worth pursuing. Once again, the whole procedure is hedged with uncertainty regarding forecasts of fuel price movements and so forth, and the method should be used with caution. This is particularly the case when an R and D project is being assessed as the forward planning is one stage more remote. The present worth of any future saving is calculated as follows. If fuel costs increase by a factor f each year, and if a discount rate of r^* per annum is applied, then the present worth of the saving in the nth year is

$$\text{PWV}_n = C(1 + f)^n/(1 + r^*)^n,$$

where C is the saving in the first year. Thus, the present worth of an installation of expected life N years is

$$\text{PWV}_N = \sum_{n=1}^{N} C(1 + f)^n/(1 + r^*)^n.$$

Note that the internal rate of return is that discount rate which would make the present worth of the plant equal to the capital investment.

It should be emphasised that the values of f and r^* that are used are not absolute ones, but assumed values over and above the overall rate of inflation, which is assumed to affect all commodities identically. There are particular uncertainties in assigning values to r^* and f, partly because r^*

is subject to changes in government attitudes to the economy and other factors, and partly because the rate of increase of fuel costs is extremely difficult to judge. In January 1979, with British inflation rates at less than 10%, and with an apparent stability in international oil prices, it would have seemed excessive to have suggested that f should be taken as 15% (representing an oil price increase of 25%). In the event, oil prices almost doubled during 1979 (92%), so that $f_{1979} = 0.8$. This introduces another complicating factor. If the heat pump system is electrically driven, and the competing system is, say, oil-fired, then the relative price movements of the two fuels must be taken into account in arriving at the value of f. During 1979, the price of electricity in the UK rose by little more than the rate of inflation, so that any effects caused by fuel price movements tended to favour the electrical heat pump rather than penalise it. In principle, this is a trend that should continue, as for a fossil-fuel power station, fuel typically constitutes only about one half of the running cost. However, principles like this seldom apply rigorously.

Because of the uncertainties introduced by this sort of effect, and because over a longer time period than one year we can expect fluctuations in relative fuel prices to average out and to approach the general level of inflation, the simplest approximation is to ignore the fuel price movement effect. This is particularly true for R and D projects that are somewhat further removed from an actual commercial decision and, in assessing the performance of the heat pump projects in its research and development programme, the EEC has adopted the three equations given below as a crude guide to economic performance (EEC 1978):

$$\text{PBP} = \Delta I/(\Delta E P_e),$$

ROR (r) is given by

$$\Delta I = \sum_{n=1}^{N} \frac{\Delta E P_e}{(1 + r)^n},$$

$$\text{PWV} = \sum_{n=1}^{N} \frac{\Delta E P_e}{(1 + r^*)^n},$$

where ΔI is the investment in achieving an energy saving, ΔE is the energy saved per year, P_e is the price of energy in the first year, N is the economic lifetime of the installation and r^* is the rate of interest on investment.

As an example, one air–water heat pump installed by the authors supplied 28 800 kWh of heat to a house for an actual electricity consumption of 9750 kWh. Of this, some 1310 kWh were supplied to the supplementary heater which was a 3 kW immersion heater in the flow line to the radiators. The seasonal coefficient of performance of the system was $28\,800/9750 = 2.95$, while that of the heat pump alone was 3.26. If we compare the heat pump with an oil-fired system, we can legitimately take fuel costs as £0.04 per kWh for electricity and £0.75 per gallon for oil (or £0.0231 per kWh assuming an efficiency of 70%). The cost of electricity in the first year is therefore £390, which is to be compared with an oil cost of £665. There is a fuel cost saving of £275.

Is this significant when compared with the extra capital expense involved in a heat pump system? At the time of writing, it seems that the heat pump will cost about £1600 and that the equivalent components of the oil-fired system (boiler, flue, tank, etc) would cost about £800. Thus, $\Delta I = £800$ and : the payback period is 800/275 = 2.9 yr; the internal rate of return is given by

$$800 = \sum_{n=1}^{N} 275/(1+r)^n,$$

and if $N = 12$ years, $r = 33\%$; the present worth, assuming no fuel price inflation, a discount rate of 10% and a life of 12 years is

$$\sum_{n=1}^{12} 275/(1+r^*)^n = £1875,$$

which is considerably greater than the ΔI of £800. Thus, the investment is well worthwhile, and becomes even more so if the discount rate is reduced, or if fuel price inflation is included. In fact, in this application, the electrically driven heat pump will still be cost effective in comparison with a traditional oil-fired system even if the capital cost is increased by 50%. Unfortunately, the same is not true if comparison is made instead with a gas-fired system at 1980 prices.

5.4.2   Energy analysis
This is an area of study that has evolved over the last few years, and which aims to elucidate the energy use patterns of a home, factory, an industry, a nation etc. Many of the techniques are carried over directly from economics, while in many applications, detailed knowledge is required of the thermodynamics of the process and of the engineering practices involved. In the case of the heat pump, the emphasis is on demonstrating the amount of the energy saving (or cost) that is derived by using the heat pump instead of the alternative system. Thus, in our example of the last section, the heat pump used 9750 kWh of electrical energy, while the heat produced is 28 800 kWh. At 70% efficiency, this would have required a quantity of oil equivalent to 41 000 kWh (890 gallons). If we assume that the electricity was generated and delivered at an efficiency of 30%, then the 9750 kWh delivered would have required the equivalent of 32 500 kWh of fuel at the power station (700 gallons). The heat pump has therefore saved the equivalent of 190 gallons of oil—or has it? Unfortunately, this type of analysis is much too simplistic as it takes no account of the fact that electricity is not generated on a large scale from the light oil that is used to fire domestic boilers. The fuel for power stations is either very heavy residual oil or low-grade coal or uranium, none of which can really be used for any other purpose. Thus, by using an electrically driven heat pump to replace a

domestic system of this capacity, 900 gallons of diesel fuel have been saved. This is where the current weakness in energy accounting lies. There is no factor included to quantify the quality of the fuel that is being consumed. How does one rate a tonne of naphtha in comparison with a quantity of uranium of the same calorific value? The naphtha has a very large number of other uses, while the uranium has no other useful application, except perhaps to reduce the volume of yacht keels (and even that is precluded by most yacht racing regulations, which prescribe lead as the material to be used).

Another factor is that, while in this case the heat pump does show a primary energy return in comparison with the oil-fired system, other alternatives might well show a better yield. For example, loft insulation, or cavity wall insulation could well reduce the primary energy consumption by the same or a greater amount. This then would require a smaller heat pump, with consequent smaller total primary energy saving, smaller fuel cost savings, and a new set of estimates to be made of capital costs, before the economics can be reassessed. Paradoxically, if insulation is increased to greater and greater extents, the heat pump eventually becomes unattractive because the total heat demand is so low that the heat pump does not seem worthwhile. Such high insulation levels are not necessarily economic, however, because of the capital cost.

One way of trying to strike a balance between the economic assessment and the crude energy analysis, is to calculate what is called an energy return, which is based on present fuel costs, and in which the discount rate is set to zero. That is, the energy cost saving over the lifetime of the machine is produced. This does nothing to answer the problem of relative fuel value, which seems set to remain an intractable problem for some time.

In some applications, such as high-temperature heat recovery, with coefficients of performance of seven and greater, heat pumps using electric drives are obviously offering both a real energy return and an economic advantage over the alternatives *now*. In some other applications, such as domestic heating using an air-source unit, the economic advantages are not so clear cut and depend on the choice of competing fuel. Interestingly, the only systems that are proposed which will compete with the electrical heat pump on a primary energy basis, are other heat pump systems such as absorption machines, and engine driven units incorporating heat recovery from the engine.

5.5   Testing of heat pumps

Experimental measurement is an essential part of any practical research programme. Additionally, verification of performance is a commercial requirement so that a manufacturer can reliably inform customers as to how his system will behave under certain conditions. For water-source, water-sink heat pumps, the test facilities required are straightforward and cause no difficulties. For air-source machines, however, the essential experimental facilities are much more complicated because of the added difficulties in stabilising both the temperature and the humidity of the air on to the evaporator. For split units, where the actual siting of the evaporator is not crucial, it is possible to build a closed cycle air conditioning loop with the evaporator coil in an exchangeable module. For packaged units, unfortunately, this option is not available and it is essential to build a larger controlled environment facility which can stabilise the temperature of the air onto the coil to $\pm\frac{1}{2}°C$ or better, and the relative humidity to better than $\pm2\%$; this stability being achieved over a range of temperatures and a range of relative humidities. For air temperatures below $0°C$, further problems occur as the measurement of air moisture levels becomes more difficult.

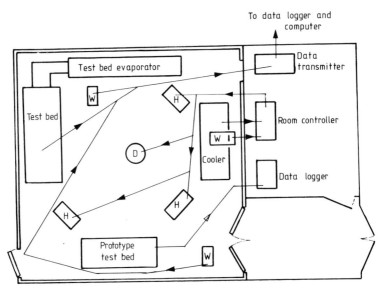

Figure 5.10 Controlled environment laboratory. W, wet and dry bulb sensors; H, heater; D, humidifier.

Figure 5.10 shows the layout of the authors' controlled environment laboratory (McMullan *et al* 1977). This provides a controlled volume $7m \times 7m \times 4m$ in which stability is achieved by using excessive cooling capacity (including the heat pump being tested) and off-setting this with proportionally controlled thyristor-chopped heaters in both the air and in a tank of water which provides water vapour to compensate for the dehumidification effect of the evaporator coils. It is important to include the heat pump being tested as part of the overall control loop because it represents such a major perturbation to the conditions in any realistic controlled environment laboratory that it would not be possible to establish the desired conditions first and then expect them to remain stable after the heat pump was started.

The air heaters are controlled by reference to platinum resistance thermometers (dry bulb), while the humidifier is controlled by reference to a differential pair of wet and dry platinum resistance thermometers. These were chosen for the control circuitry as they give directly signals large enough for controlling electronic circuitry, and their resistance is proportional to temperature. By contrast, thermocouples give only small voltages, and thermistors are logarithmic. They also require no reference junction, and operate at subzero temperatures without a change in signal voltage polarity.

Having achieved stable and known conditions, it is essential to decide the purpose of the measurements being made, the number of sampling points that are required and the accuracy. For example, it is fairly easy to measure temperatures to within $\pm\frac{1}{2}°C$. However, in a water-sink heat pump with a flow temperature of $50°C$ and a return temperature of $40°C$, this error will lead to an uncertainty of $\pm7\%$ in the measurement of heat output, which is unacceptable. Thus, much more accurate temperature measurement is required.

If the experimental tests are merely to establish performance, then fairly simple measurements are all that are required: environmental conditions, accurate measurement of the power supplied, the flow and return temperatures and mass flow rate of the heat distribution medium. For more detailed research, more complex measurements are necessary to determine state points at different points in the cycle and so forth. If long data analysis runs are to be attempted then a microprocessor based data logging system with disk storage or direct access to a larger computer is invaluable, as the amount of data collected rapidly exceeds that which is manageable—or even useful.

A much more thorny problem than that of data collection, however, is the establishment of acceptable and believeable reporting standards and criteria. At the time of writing, there is no recognised set of reporting standards that allows either potential customers or other research workers to understand unequivocally the results reported in any one investigation. There has been widespread recognition that there is a problem, but there has been no agreement as to any course of action to deal with it. There has been at least one draft specification produced by CIB (1977) supposedly setting a European standard heat pump specification, but the document has so many spaces left for individual national legislation or practice that one is left with what is primarily a recommended reporting style.

For research work, a very large gap remains, and the only real guide is set by the EEC (1978) guidelines for presenting final reports on research contracts. This asks for the performance of the heat pump to be presented under design conditions with the following being reported specifically:

(a) source temperature;
(b) sink temperature (outlet temperature);
(c) capacity of the heat pump; and
(d) performance, (i) mechanically driven—COP and (ii) thermally driven—PER.

The same data should also be given under off-design conditions, both in graphical form and in tabular form to show

(a) the influence of the temperature of the heat sink and source on peformance; and
(b) the influence of the load.

For domestic applications, the seasonal performance should also be given and defrosting load should be included in the figures. In particular, the fan or pump power needed to transport the heat source medium over the evaporator should be included. Also required is an estimate of the exergetic efficiency $\eta_{exerg} = COP_{real}/COP_{Carnot}$ both on a system basis and on the internal refrigerant cycle basis, neglecting superheat.

These suggestions go some of the way towards standardising the reporting of heat pump performance, but there are still a very large number of reporting problems because of the diversity of operating conditions and the diversity of systems. There seems likely to be no standard agreement until one begins to emerge through the normal reporting processes. For domestic units, however, and for air-source units in particular, it would be of great benefit if some standardised air-source conditions such as 5 °C and 80% relative humidity could *always* be reported regardless of the condenser conditions as these will vary according to application. In this way, at least some comparison between different units could be established. The relative humidity parameter is important because, as was discussed in Chapter 2, latent heat pick-up from the atmosphere is an important factor and can seriously influence reported performance data.

5.6 Future prospects

In the early 1970s the primary impediment to the widespread adoption of heat pumps was the fact that fossil fuels were so cheap that there was no incentive. In the early 1980s this is not the case, and we can expect heat pumps to become more popular as time goes by. The reasons why they have not become instantly popular with the increase in fuel prices are not hard to find: high initial cost, conservatism and scepticism from both customers and manufacturers, a vain hope that the energy problem would disappear, and administrative or semitechnical obstacles.

The administrative obstacles can be typified by the trade codes in West Germany which specify what work can be done by what craftsmen, with overstepping of the limits being treated as an infringement of the law. Rigorous adherence to this code would require up to four different trades-men to be involved in a heat pump installation (refrigeration technician, heating engineer, electrician and builder). Another example, again from West Germany, involves the chimney sweep, who is by law an adviser to the public on all questions relating to space heating. What are the problems imposed by a heating system that has no chimney and hence no chimney sweep?

On the semitechnical side, the United Kingdom has its share of complexities, such as electricity boards who will not permit the connection of a motor larger than one or two horsepower to a single-phase electricity supply, yet will not install three-phase domestic supplies.

It is probably safe to say now that only a tiny minority still believe that the energy crisis is a fabrication of some unholy alliance between the oil producers and the oil companies. Thus, this obstacle is now removed. Conservatism and scepticism are still around, however, though interest in heat pumps is very greatly increased. The conservatism tends to make customers continue to buy the old well established heating systems, with an increasing preference for solid fuel rather than oil. The scepticism stems rather more from a misplaced optimism about other possibilities rather than true scepticism about the heat pump. Here the problem is inadequate information, and it is unfortunately true that the more glamorous 'alternative energy sources' are far more frequently the subject of extensive magazine articles than heat pumps, despite the fact that the heat pump has a proven ability to conserve fuel supplies now, while the alternatives do not as yet.

However, the real impediment is the high initial cost; yet another fruitful area for budding heat pump research workers. This should not be an undue obstacle to an industrial customer, who has access to the necessary accounting techniques to show whether or not his investment is worthwhile, but private citizens do not analyse their domestic accounts in this way. Industry is recognising the heat recovery potential of the heat pump, and momentum is increasing for the installation of new equipment. On the domestic side, however, the market is still primarily a prestige one—which allows sales to be made at a relatively high capital cost. Only the development of a larger market, with corresponding decreases in unit costs, can fully rectify this problem, but it is important to reduce initial costs as much as possible *now* to encourage rapid expansion.

APPENDIX

Listing of the refrigerant properties program suite, refrigerant data, and sample of output. The programs are written in ICL 1900 Extended Fortran but should need only FORMAT changes on transfer to another computer.

```
 1   C        PROGRAM TO CALCULATE THE STATE POINT PROPERTIES OF A REFRIGERANT.
 2   C        THE REFRIGERANT CAN BE ONE OF R11, R12, R21, R22, R114, R502.
 3   C        THE REQUIRED CONDITIONS ARE PRESSURE AND ONE OTHER PROPERTY
 4   C        AS REQUESTED BY THE PROGRAM.
 5   C
 6            MASTER RPROP
 7            WRITE(2,200)
 8   200  FORMAT(///5X,'REFRIGERANT PROPERTIES PROGRAM'/
 9        *5X,30('*')//
10        *' REFRIGERANT NO ? '/)
11            READ(1,100)NR
12   100  FORMAT(I0)
13            CALL RPDATA(NR)
14            WRITE(2,201)
15   201  FORMAT(/' UNITS ?  TYPE 0 FOR B.TH.UNITS'/
16        *15X,'1 FOR S.I. UNITS'/)
17            READ(1,100)IU
18   10   WRITE(2,202)
19   202  FORMAT(//' PRESSURE ? (= 0 TO STOP PROGRAM)'/)
20            READ(1,101)P
21   101  FORMAT(F0.0)
22            IF(P.LE.0.0)PAUSE
23            WRITE(2,203)
24   203  FORMAT(/' SECOND PARAMETER ?  TYPE 1 FOR TEMPERATURE'/
25        *26X,'2 FOR SPECIFIC VOLUME'/
26        *26X,'3 FOR SPECIFIC ENTHALPY'/
27        *26X,'4 FOR SPECIFIC ENTROPY'/
28        *26X,'5 FOR QUALITY'/)
29            READ(1,100)I
30            GOTO(20,30,40,50,60),I
31   20   WRITE(2,204)
32   204  FORMAT(/' TEMPERATURE ?'/)
33            READ(1,101)T
34            TSA=TSAT(NR,T,IU)
35            IF(ABS(T-TSA).GT.0.0001)GOTO70
36            WRITE(2,205)
37   205  FORMAT(/' QUALITY ?'/)
38            READ(1,101)X
39            GOTO70
40   30   WRITE(2,206)
41   206  FORMAT(/' SPEC VOLUME ?'/)
42            READ(1,101)V
43            GOTO70
44   40   WRITE(2,207)
45   207  FORMAT(/' SPEC ENTHALPY ?'/)
46            READ(1,101)H
47            GOTO70
48   50   WRITE(2,208)
```

```
49       208 FORMAT(/' SPEC ENTROPY ?'/)
50           READ(1,101)S
51           GOTO70
52        60 WRITE(2,205)
53           READ(1,101)X
54        70 CALL REFPRP(NR,T,P,V,H,S,X,I,IU)
55           WRITE(2,210)NR,T,P,V,H,S,X
56       210 FORMAT(//' R',I3//
57          *' T =',F7.2/' P =',F10.5/' V =',F10.5/' H =',F10.5/
58          *' S =',F10.5/' X =',F10.5/)
59           GOTO10
60           END


61           SUBROUTINE RPDATA(NR)
62     C   THIS SUBROUTINE READS IN THE  CONSTANTS AND COEFFICIENTS
63     C   USED FOR CALCULATING THE THERMODYNAMIC PROPERTIES OF A REFRIGERANT.
64     C   THIS SUBROUTINE SHOULD BE CALLED ONCE ONLY, AT THE START OF ANY
65     C   PROGRAM REQUIRING REFRIGERANT PROPERTIES, OR WHEN A CHANGE OF
66     C   REFRIGERANT IS REQUIRED.
67     C   THE CONSTANTS AND COEFFICIENTS APPROPRIATE TO VARIOUS REFRIGERANTS
68     C   MAY BE FOUND IN THE PAPER BY DOWNING : "REFRIGERANT EQUATIONS"
69     C   (TRANSACTIONS ASHRAE : PAPER NO. 2313)
70     C
71           REAL J,K,LE10,L10E
72           COMMON/CVHSJ/ACV,BCV,CCV,DCV,ECV,FCV,X,Y,J,LE10,L10E
73           COMMON/STATEQ/R,B,A2,B2,C2,A3,B3,C3,A4,B4,C4,A5,B5,C5,A6,B6,C6,
74          *K,ALPHA,CPR
75           COMMON/TMF/TC,TFR
76           COMMON/LIQ/AL,BL,CL,DL,EL,FL,GL
77           COMMON/SAT/AVP,BVP,CVP,DVP,EVP,FVP
78           COMMON/TST/PCRIT,AA,BB
79           COMMON/CKMS/ACPL(7),ACPV(7),AKL(7),AKV(7),AML(7),AMV(7),
80          *TMINL,TMAXL,TMINV
81           J=0.185053
82           LE10=2.302585093
83           L10E=0.4342944819
84           NRR=NR
85     C   REFRIGERANT NUMBER
86     C
87         1 READ(5,200)NR
88     C   EQUATION OF STATE CONSTANTS
89     C
90           READ(5,100)R,B,A2,B2,C2,A3,B3,C3,A4,B4,C4,A5,B5,C5,A6,B6,C6,
91          *K,ALPHA,CPR,TC,TFR
92     C   LIQUID DENSITY CONSTANTS
93     C
94           READ(5,100)AL,BL,CL,DL,EL,FL,GL
95     C   VAPOUR PRESSURE CONSTANTS
96     C
97           READ(5,100)AVP,BVP,CVP,DVP,EVP,FVP
98     C   SPECIFIC HEAT AT CONSTANT VOLUME CONSTANTS
99     C
100          READ(5,100)ACV,BCV,CCV,DCV,ECV,FCV
101    C   VAPOUR ENTHALPY AND ENTROPY CONSTANTS
102    C
103          READ(5,100)X,Y
104    C   CRITICAL PRESSURE
105    C
106          READ(5,100)PCRIT
107    C   COEFFICIENTS FOR INITIAL ESTIMATE OF 'TSAT'
108    C   ( TSAT=AA*ALOG10(P)+BB )
109    C
110          READ(5,100)AA,BB
111    C   THERMOPHYSICAL PROPERTY CONSTANTS
112    C
113          READ(5,100)ACPL,ACPV,AKL,AKV,AML,AMV,TMINL,TMAXL,TMINV
114          IF(NR.NE.NRR)GOTO1
115          RETURN
116      100 FORMAT(G0.0)
117      200 FORMAT(I0)
118          END


119          SUBROUTINE REFPRP(NR,T,P,V,H,S,X,I,IU)
120    C
121    C   DETERMINES THE THERMODYNAMIC PROPERTIES OF A REFRIGERANT,
122    C   GIVEN PRESSURE AND ONE OTHER PARAMETER.
123    C   THE PHASE OF THE REFRIGERANT IS DETERMINED AUTOMATICALLY.
```

133

```
124    C
125    C           IU = 0 : B.TH.UNITS
126    C                1 : S.I. UNITS
127    C
128    C            I = 1 : SECOND PARAMETER IS TEMPERATURE
129    C                2 :                      SPECIFIC VOLUME
130    C                3 :                      SPECIFIC ENTHALPY
131    C                4 :                      SPECIFIC ENTROPY
132    C                5 :                      THERMODYNAMIC QUALITY
133    C
134          TOL=0.0001
135          TSA=TSAT(NR,P,IU)
136          CALL SATPRP(NR,TSA,P,VF,VG,HF,HFG,HG,SF,SG,IU)
137          J=2
138          GOTO(10,20,30,40,50),I
139          GOTO600
140     10 IF((TSA-T).GT.TOL)GOTO12
141          IF((T-TSA).GT.TOL)GOTO14
142          V=VF+(VG-VF)*X
143          H=HF+HFG*X
144          S=SF+(SG-SF)*X
145          GOTO500
146     12 CALL SCOL(NR,T,P,V,H,S,IU)
147          GOTO16
148     14 CALL VAPOR(NR,T,P,V,H,S,IU)
149     16 X=(H-HF)/HFG
150          GOTO500
151     20 Y=V
152          IF(((VF-V)/V).GT.TOL)J=1
153          IF(((V-VG)/V).GT.TOL)J=3
154          GOTO100
155     30 Y=H
156          IF(((HF-H)/H).GT.TOL)J=1
157          IF(((H-HG)/H).GT.TOL)J=3
158          GOTO100
159     40 Y=S
160          IF(((SF-S)/S).GT.TOL)J=1
161          IF(((S-SG)/S).GT.TOL)J=3
162          GOTO100
163     50 Y=X
164          IF(X.LT.0.0)J=1
165          IF(X.GT.1.0)J=3
166    C
167    C  CALCULATIONS FOR TWO-PHASE CASE
168    C
169    100 IF(J.NE.2)GOTO200
170          T=TSA
171          GOTO(10,120,130,140,150),I
172    120 X=(V-VF)/(VG-VF)
173          H=HF+HFG*X
174          S=SF+(SG-SF)*X
175          GOTO500
176    130 X=(H-HF)/HFG
177          V=VF+(VG-VF)*X
178          S=SF+(SG-SF)*X
179          GOTO500
180    140 X=(S-SF)/(SG-SF)
181          V=VF+(VG-VF)*X
182          H=HF+HFG*X
183          GOTO500
184    150 V=VF+(VG-VF)*X
185          H=HF+HFG*X
186          S=SF+(SG-SF)*X
187          GOTO500
188    C
189    C  ITERATION LOOP FOR SUBCOOLED AND SUPERHEATED CASES
190    C
191    200 T1=TSA
192          DT=0.000001
193          DO 300 L=1,20
194          T11=T1+DT
195          GOTO(210,100,220),J
196    210 CALL SCOL(NR,T1,P,V1,H1,S1,IU)
197          CALL SCOL(NR,T11,P,V11,H11,S11,IU)
198          GOTO230
199    220 CALL VAPOR(NR,T1,P,V1,H1,S1,IU)
200          CALL VAPOR(NR,T11,P,V11,H11,S11,IU)
201    230 GOTO(10,232,233,234,235),I
202    232 Y1=V1
```

134

```
203          Y11=V11
204          GOTO240
205     233  Y1=H1
206          Y11=H11
207          GOTO240
208     234  Y1=S1
209          Y11=S11
210          GOTO240
211     235  Y1=(H1-HF)/HFG
212          Y11=(H11-HF)/HFG
213     240  IF((ABS(Y-Y1)/Y).LE.TOL)GOTO400
214          T2=T1+DT*(Y-Y1)/(Y11-Y1)
215     300  T1=T2
216          WRITE(2,1001)
217     400  T=T1
218          IF(I.NE.2)V=V1
219          IF(I.NE.3)H=H1
220          IF(I.NE.4)S=S1
221          IF(I.NE.5)X=(H1-HF)/HFG
222     500  RETURN
223     600  WRITE(2,1002)I
224          PAUSE 'REFPRP'
225    1001  FORMAT(/' *** WARNING *** REFPRP : NOT CONVERGED'/)
226    1002  FORMAT(/' ***   ERROR   *** REFPRP : I =',I3/)
227          END


228          SUBROUTINE VAPOR2(NR,T,P,V,H,S,I,IU)
229   C
230   C  DETERMINES THE THERMODYNAMIC PROPERTIES OF SUPERHEATED REFRIGERANT
231   C  GIVEN PRESSURE AND ONE OTHER PROPERTY
232   C           T = TEMPERATURE       DEG C    (DEG F)
233   C           P = ABSOLUTE PRESSURE  BAR     (P.S.I.A.)
234   C           V = SPECIFIC VOLUME    CU M/KG (CU FT/LB)
235   C           H = SPECIFIC ENTHALPY  KJ/KG   (BTU/LB)
236   C           S = SPECIFIC ENTROPY   KJ/KG DEG C (BTU/LB DEG F)
237   C           I=1 : 2ND INPUT IS TEMPERATURE
238   C           I=2 : 2ND INPUT IS SPEC VOLUME
239   C           I=3 : 2ND INPUT IS ENTHALPY
240   C           I=4 : 2ND INPUT IS ENTROPY
241   C           IU=0 - B.TH. UNITS   IU=1 - S.I. UNITS
242   C
243          GOTO(10,20,30,40),I
244          GOTO600
245     10   CALL VAPOR(NR,T,P,V,H,S,IU)
246          GOTO500
247     20   Y=V
248          TOL=0.00000001
249          IF(IU.EQ.0)TOL=0.000001
250          GOTO100
251     30   Y=H
252          TOL=0.00001
253          GOTO100
254     40   Y=S
255          TOL=0.000001
256   C  ITERATE TO FIND PROPERTIES
257   C
258     100  T1=TSAT(NR,P,IU)
259          DT=0.000001
260          DO 200 L=1,20
261          T11=T1+DT
262          CALL VAPOR(NR,T1,P,V1,H1,S1,IU)
263          CALL VAPOR(NR,T11,P,V11,H11,S11,IU)
264          GOTO(10,120,130,140),I
265     120  Y1=V1
266          Y11=V11
267          GOTO150
268     130  Y1=H1
269          Y11=H11
270          GOTO150
271     140  Y1=S1
272          Y11=S11
273     150  CONTINUE
274          IF(ABS(Y-Y1).LE.TOL)GOTO300
275          T2=T1+DT*(Y-Y1)/(Y11-Y1)
276     200  T1=T2
277          WRITE(2,1001)
278     300  T=T1
279          IF(I.NE.2)V=V1
```

135

```
280          IF(I.NE.3)H=H1
281          IF(I.NE.4)S=S1
282      500 RETURN
283      600 WRITE(2,1002)I
284          PAUSE 'VAPOR2'
285     1001 FORMAT(/' *** WARNING *** VAPOR2 : NOT CONVERGED'/)
286     1002 FORMAT(/' ***  ERROR   *** VAPOR2 : I =',I3/)
287          END


288          SUBROUTINE VAPOR(NR,TF,PPSIA,VVAP,HVAP,SVAP,IU)
289    C
290    C  DETERMINES THE THERMODYNAMIC PROPERTIES OF SUPERHEATED
291    C  REFRIGERANT, GIVEN TEMPERATURE AND PRESSURE
292    C
293    C          TF   = TEMPERATURE       DEG C   (DEG F)
294    C          PPSIA= ABSOLUTE PRESSURE BAR     (P.S.I.A.)
295    C          VVAP = SPECIFIC VOLUME   CU M/KG (CU FT/LB)
296    C          HVAP = SPECIFIC ENTHALPY KJ/KG   (BTU/LB)
297    C          SVAP = SPECIFIC ENTROPY  KJ/KG DEG C (BTU/LB DEG F)
298    C
299    C          IU=0 - B.TH. UNITS   IU=1 - S.I. UNITS
300    C
301          REAL J,K,KTDTC,LE10,L10E
302          COMMON/CVHSJ/ACV,BCV,CCV,DCV,ECV,FCV,X,Y,J,LE10,L10E
303          COMMON/STATEQ/R,B,A2,B2,C2,A3,B3,C3,A4,B4,C4,A5,B5,C5,A6,B6,C6,
304         *K,ALPHA,CPR
305          COMMON/TMP/TC,TFR
306          IF(IU.LT.1)GOTO100
307          TF=TF*1.8+32.0
308          PPSIA=PPSIA*14.5
309      100 CONTINUE
310    C
311    C  CONVERT 'TF' TO ABSOLUTE TEMP (DEG R) AND CHECK IF ABOVE ZERO
312    C
313          T=TF+TFR
314          IF(T.LE.0.0) GO TO 902
315    C
316    C  CHECK IF TEMP.GE.SATURATION TEMP
317    C
318          TFSAT=TSAT(NR,PPSIA,0)
319          IF(TF.LT.(TFSAT-0.0001)) GO TO 903
320    C
321    C  CHECK IF PRESSURE ABOVE ZERO
322    C
323          IF(PPSIA.LE.0.0) GO TO 904
324    C
325    C  CALCULATE SPECIFIC VOLUME 'VVAP'
326    C
327          VVAP=SPVOL(NR,TF,PPSIA,0)
328    C
329    C  CALCULATE SPEC ENTHALPY 'HVAP' AND SPEC ENTROPY 'SVAP'
330    C
331          T2=T*T
332          T3=T2*T
333          T4=T3*T
334          VR=VVAP-B
335          VR2=2.0*VR*VR
336          VR3=1.5*VR2*VR
337          VR4=VR2*VR2
338          KTDTC=K*T/TC
339          EKTDTC=EXP(-KTDTC)
340          EMAV=EXP(-ALPHA*VVAP)
341          H1=ACV*T+BCV*T2/2.+CCV*T3/3.+DCV*T4/4.-FCV/T
342          H2=J*PPSIA*VVAP
343          H3=A2/VR+A3/VR2+A4/VR3+A5/VR4
344          H4=C2/VR+C3/VR2+C4/VR3+C5/VR4
345          S1=ACV*ALOG(T)+BCV*T+CCV*T2/2.+DCV*T3/3.-FCV/
346         1(2.*T2)
347          S2=J*R*ALOG(VR)
348          S3=B2/VR+B3/VR2+B4/VR3+B5/VR4
349          S4=H4
350          IF(ABS(ALPHA).LE.1.0E-20)GOTO6
351          IF(ABS(CPR).GT.1.0E-20)GOTO5
352        4 H3=H3+A6/ALPHA*EMAV
353          S3=S3+B6/ALPHA*EMAV
354          GO TO 6
355        5 H0=1./ALPHA*(EMAV-CPR*ALOG(1.+EMAV/CPR))
356          H3=H3+A6*H0
```

```
357          H4=H4+C6*H0
358          S3=S3+B6*H0
359          S4=S4+C6*H0
360        6 HVAP=H1+H2+J*H3+J*EKTDTC*(1.+KTDTC)*H4+X
361          SVAP=S1+S2-J*S3+J*EKTDTC*K/TC*S4+Y
362          IF(IU.LT.1)GOTO200
363          TF=(TF-32.0)/1.8
364          PPSIA=PPSIA/14.5
365          VVAP=VVAP*0.0624219
366          HVAP=HVAP*2.326
367          SVAP=SVAP*4.1868
368      200 RETURN
369      902 WRITE(2,1002)
370          GOTO999
371      903 WRITE(2,1003)
372          GOTO999
373      904 WRITE(2,1004)
374      999 PAUSE 'VAPOR'
375     1002 FORMAT(/' ***   ERROR   *** VAPOR : TEMP.LE.ZERO'/)
376     1003 FORMAT(/' ***   ERROR   *** VAPOR : TEMP.LT.SAT TEMP'/)
377     1004 FORMAT(/' ***   ERROR   *** VAPOR : PRESSURE.LE.ZERO'/)
378          END


379          SUBROUTINE SCOL(NR,TF,PPSIA,VA,HA,SA,IU)
380    C
381    C  DETERMINES THE THERMODYNAMIC PROPERTIES OF SUBCOOLED LIQUID
382    C  REFRIGERANT, GIVEN TEMPERATURE AND PRESSURE
383    C
384    C          TF   = TEMPERATURE          DEG C   (DEG F)
385    C          PPSIA= ABSOLUTE PRESSURE    BAR     (P.S.I.A.)
386    C          VA   = SPECIFIC VOLUME      CU M/KG (CU FT/LB)
387    C          HA   = SPECIFIC ENTHALPY    KJ/KG   (BTU/LB)
388    C          SA   = SPECIFIC ENTROPY     KJ/KG DEG C (BTU/LB DEG F)
389    C
390    C          IU = 0 : B.TH. UNITS   IU = 1 : S.I. UNITS
391    C
392          REAL HA,HL,HG,HLG
393          IF(IU.LT.1)GOTO100
394          TF=TF*1.8+32.0
395          PPSIA=PPSIA*14.5
396      100 CONTINUE
397          T=TSAT(NR,PPSIA,0)
398          DELT=T-TF
399    C
400    C  CHECK IF TEMP.LE.SATURATION TEMP
401    C
402          IF(DELT.LT.-0.0001)GOTO901
403          CALL SATPRP(NR,T,PSA,VL,VG,HL,HLG,HG,SL,SG,0)
404          VA=VL
405          IF(ABS(T+40.0).LE.1.0E-06)GOTO130
406          HA=HL*(1.0-DELT/(T+40.))
407          GOTO 140
408      130 HA=HL
409      140 IF(ABS(TF+40.0).LE.1.0E-06)GOTO150
410          SA=SL+(HL/(TF+40.))*ALOG((TF+459.7)/(T+459.7))
411          GOTO160
412      150 SA=SL
413      160 CONTINUE
414          IF(IU.LT.1)GOTO200
415          TF=(TF-32.0)/1.8
416          PPSIA=PPSIA/14.5
417          VA=VA*0.0624219
418          HA=HA*2.326
419          SA=SA*4.1868
420      200 RETURN
421      901 WRITE(2,1001)
422      999 PAUSE 'SCOL'
423     1001 FORMAT(/' ***   ERROR   *** SCOL : TEMP.GT.SAT TEMP'/)
424          END


425          SUBROUTINE SATPRP(NR,TF,PSA,VF,VG,HF,HFG,HG,SF,SG,IU)
426    C
427    C  DETERMINES THE SATURATION THERMODYNAMIC PROPERTIES OF
428    C  A REFRIGERANT GIVEN THE SATURATION TEMPERATURE
429    C          TF   = TEMPERATURE          DEG C   (DEG F)
430    C          PSA  = SATURATION PRESSURE  BAR     (P.S.I.A.)
431    C          VF   = SPEC VOL OF SAT LIQUID  CU M/KG (CU FT/LB)
```

```
432  C              VG   = SPEC VOL OF SAT VAPOUR      CU M/KG  (CU FT/LB)
433  C              HF   = SPEC ENTH OF SAT LIQ        KJ/KG    (BTU/LB)
434  C              HFG  = LATENT HEAT OF VAPORISATION KJ/KG    (BTU/LB)
435  C              HG   = SPEC ENTH OF SAT VAPOUR     KJ/KG    (BTU/LB)
436  C              SF   = SPEC ENTROPY SAT LIQUID     KJ/KG DEG C  (BTU/LB DEG F)
437  C              SG   = SPEC ENTROPY SAT VAPOUR     KJ/KG DEG C  (BTU/LB DEG F)
438  C              IU=0 - B.TH. UNITS    IU=1 - S.I. UNITS
439  C
440          REAL J,K,KTDTC,LE10,L10E
441          COMMON/LIQ/AL,BL,CL,DL,EL,FL,GL
442          COMMON/CVHSJ/ACV,BCV,CCV,DCV,ECV,FCV,X,Y,J,LE10,L10E
443          COMMON/STATEQ/R,B,A2,B2,C2,A3,B3,C3,A4,B4,C4,A5,B5,C5,A6,B6,C6,
444         *K,ALPHA,CPR
445          COMMON/TMP/TC,TFR
446          COMMON/SAT/AVP,BVP,CVP,DVP,EVP,FVP
447          IF(IU.LT.1)GOTO100
448          TF=TF*1.8+32.0
449     100 CONTINUE
450  C
451  C  CONVERT 'TF' TO ABSOLUTE TEMP (DEG R) AND CHECK IF ABOVE ZERO
452  C
453          T=TF+TFR
454          IF(T.LE.0.0) GO TO 902
455  C
456  C  CHECK IF TEMP.LE.CRITICAL TEMP
457  C
458          IF(T.GT.TC) GO TO 903
459  C
460  C  CALCULATE PRESSURE
461  C
462      11 PSA=10.**(AVP+BVP/T+CVP*ALOG10(T)+DVP*T+EVP*
463         1((FVP-T)/T)*ALOG10(ABS(FVP-T)))
464  C
465  C  CALCULATE SPECIFIC VOLUME OF SAT VAPOUR 'VG'
466  C
467      12 VG=SPVOL(NR,TF,PSA,0)
468  C
469  C  CALCULATE SPECIFIC VOLUME OF SAT LIQUID 'VF'
470  C
471       2 IF(NR.EQ.21.OR.NR.EQ.113)GOTO20
472          TR1=1.-T/TC
473          VF=1./(AL+BL*TR1**(1./3.)+CL*TR1**(2./3.)+DL*TR1+
474         *EL*TR1**(4./3.)+FL*TR1**0.5+GL*TR1*TR1)
475          GOTO21
476      20 VF=1.0/(AL+BL*T+CL*T*T)
477      21 CONTINUE
478  C
479  C  CALCULATE LATENT HEAT 'HFG' BY CLAUSIUS CLAPEYRON EQUATION
480  C
481          IF(ABS(EVP).LE.1.0E-20)GOTO30
482          HFG=(VG-VF)*PSA*LE10*(-BVP/T+CVP/LE10+DVP*T-
483         1EVP*(L10E+FVP*ALOG10(FVP-T)/T))*J
484          GOTO31
485      30 HFG=(VG-VF)*PSA*LE10*(-BVP/T+CVP/LE10+DVP*T)*J
486      31 CONTINUE
487          SFG=HFG/T
488  C
489  C  CALCULATE SPEC ENTHALPY 'HG' AND SPEC ENTROPY 'SG' (SAT VAPOUR)
490  C
491          T2=T*T
492          T3=T2*T
493          T4=T3*T
494          VR=VG-B
495          VR2=2.0*VR*VR
496          VR3=1.5*VR2*VR
497          VR4=VR2*VR2
498          KTDTC=K*T/TC
499          EKTDTC=EXP(-KTDTC)
500          EMAV=EXP(-ALPHA*VG)
501          H1=ACV*T+BCV*T2/2.+CCV*T3/3.+DCV*T4/4.-FCV/T
502          H2=J*PSA*VG
503          H3=A2/VR+A3/VR2+A4/VR3+A5/VR4
504          H4=C2/VR+C3/VR2+C4/VR3+C5/VR4
505          S1=ACV*ALOG(T)+BCV*T+CCV*T2/2.+DCV*T3/3.-FCV/
506         1(2.*T2)
507          S2=J*R*ALOG(VR)
508          S3=B2/VR+B3/VR2+B4/VR3+B5/VR4
509          S4=H4
510          IF(ABS(ALPHA).LE.1.0E-20)GOTO6
```

138

```
511          IF(ABS(CPR).GT.1.0E-20)GOTO5
512        4 H3=H3+A6/ALPHA*EMAV
513          S3=S3+B6/ALPHA*EMAV
514          GO TO 6
515        5 H0=1./ALPHA*(EMAV-CPR*ALOG(1.+EMAV/CPR))
516          H3=H3+A6*H0
517          H4=H4+C6*H0
518          S3=S3+B6*H0
519          S4=S4+C6*H0
520        6 HG=H1+H2+J*H3+J*EKTDTC*(1.+KTDTC)*H4+X
521          SG=S1+S2-J*S3+J*EKTDTC*K/TC*S4+Y
522   C
523   C  CALCULATE SPEC ENTHALPY 'HF' AND SPEC ENTROPY 'SF' (SAT LIQUID)
524   C
525          HF=HG-HFG
526          SF=SG-SFG
527          IF(IU.LT.1)GOTO200
528          TF=(TF-32.0)/1.8
529          PSA=PSA/14.5
530          VF=VF*0.0624219
531          VG=VG*0.0624219
532          HF=HF*2.326
533          HFG=HFG*2.326
534          HG=HG*2.326
535          SF=SF*4.1868
536          SG=SG*4.1868
537      200 RETURN
538      902 WRITE(2,1002)
539          GOTO999
540      903 WRITE(2,1003)
541      999 PAUSE 'SATPRP'
542     1002 FORMAT(/' ***   ERROR   *** SATPRP : TEMP.LE.ZERO'/)
543     1003 FORMAT(/' ***   ERROR   *** SATPRP : TEMP.GT.CRIT TEMP'/)
544          END


545          FUNCTION SPVOL(NR,TF,PPSIA,IU)
546   C
547   C  DETERMINES THE SPECIFIC VOLUME OF SUPERHEATED REFRIGERANT
548   C  GIVEN THE TEMPERATURE AND PRESSURE
549   C
550   C          TF   = TEMPERATURE        DEG C   (DEG F)
551   C          PPSIA= ABSOLUTE PRESSURE  BAR     (P.S.I.A.)
552   C          SPVOL= SPECIFIC VOLUME    CU M/KG (CU FT/LB)
553   C
554   C          IU=0 : B.TH. UNITS   IU=1 : S.I. UNITS
555   C
556          REAL K
557          COMMON/STATEQ/R,B,A2,B2,C2,A3,B3,C3,A4,B4,C4,A5,B5,C5,A6,B6,C6,
558         *K,ALPHA,CPR
559          COMMON/TMP/TC,TFR
560          IF(IU.LT.1)GOTO100
561          TF=TF*1.8+32.0
562          PPSIA=PPSIA*14.5
563      100 CONTINUE
564   C
565   C  CONVERT 'TF' TO ABSOLUTE TEMP (DEG R) AND CHECK IF ABOVE ZERO
566   C
567          T=TF+TFR
568          IF(T.LE.0.0) GO TO 902
569   C
570   C  CHECK IF TEMP.GE.SATURATION TEMP
571   C
572          TFSAT=TSAT(NR,PPSIA,0)
573          IF(TF.LT.(TFSAT-0.0001)) GO TO 903
574   C
575   C  CHECK IF PRESSURE ABOVE ZERO
576   C
577           IF(PPSIA.LE.1.0E-20) GO TO 904
578   C
579   C  CALCULATE CONSTANTS
580   C
581           ES0=EXP(-K*T/TC)
582           ES1=PPSIA
583           ES2=R*T
584           ES3=A2+B2*T+C2*ES0
585           ES4=A3+B3*T+C3*ES0
586           ES5=A4+B4*T+C4*ES0
587           ES6=A5+B5*T+C5*ES0
```

139

```
588          ES7=A6+B6*T+C6*ES0
589          ES32=2.*ES3
590          ES43=3.*ES4
591          ES54=4.*ES5
592          ES65=5.*ES6
593   C
594   C   COMPUTE INITIAL VALUE OF 'V' FROM GAS LAW
595   C
596          VN=R*T/PPSIA
597   C
598   C   COMPUTE 'V' TO WITHIN (1.E-08*V) BY ITERATION
599   C
600          DO 10 L=1,50
601          V=VN
602          V2=V*V
603          V3=V2*V
604          V4=V3*V
605          V5=V4*V
606          V6=V5*V
607          EMAV=EXP(-ALPHA*(V+B))
608          IF(ABS(CPR).GT.1.0E-20)GOTO3
609        2 F=ES1-ES2/V-ES3/V2-ES4/V3-ES5/V4-ES6/V5-ES7*EMAV
610          FV=ES2/V2+ES32/V3+ES43/V4+ES54/V5+ES65/V6+ES7*ALPHA*EMAV
611          GO TO 4
612        3 EM2AV=EMAV*EMAV
613          F=ES1-ES2/V-ES3/V2-ES4/V3-ES5/V4-ES6/V5-ES7*EM2AV/(EMAV+CPR)
614          FV=ES2/V2+ES32/V3+ES43/V4+ES54/V5+ES65/V6+ES7*EM2AV*ALPHA
615         1*(EMAV+2.*CPR)/(EMAV+CPR)**2
616        4 FFV=F/FV
617          IF(ABS(FFV).GE.VN) FFV=FFV/5.
618          VN=V-FFV
619          IF(ABS((VN-V)/V).LE.1.E-08) GO TO 20
620       10 CONTINUE
621          WRITE(2,1000)
622       20 SPVOL=VN+B
623          IF(IU.LT.1)GOTO200
624          TF=(TF-32.0)/1.8
625          PPSIA=PPSIA/14.5
626          SPVOL=SPVOL*0.0624219
627      200 RETURN
628      902 WRITE(2,1002)
629          GOTO999
630      903 WRITE(2,1003)
631          GOTO999
632      904 WRITE(2,1004)
633      999 PAUSE 'SPVOL'
634     1000 FORMAT(/' *** WARNING *** SPVOL : NOT CONVERGED'/)
635     1002 FORMAT(/' ***   ERROR   *** SPVOL : TEMP.LE.ZERO'/)
636     1003 FORMAT(/' ***   ERROR   *** SPVOL : TEMP.LT.SAT TEMP'/)
637     1004 FORMAT(/' ***   ERROR   *** SPVOL : PRESSURE.LE.ZERO'/)
638          END


639          FUNCTION TSAT(NR,PSA,IU)
640   C
641   C   DETERMINES SATURATION TEMPERATURE OF A REFRIGERANT
642   C   GIVEN SATURATION PRESSURE
643   C
644   C          PSA  = ABSOLUTE PRESSURE  BAR   (P.S.I.A.)
645   C          TSAT = TEMPERATURE        DEG C (DEG F)
646   C
647   C          IU=0 - B.TH. UNITS   IU=1 - S.I. UNITS
648   C
649          REAL LE10
650          COMMON/SAT/AVP,BVP,CVP,DVP,EVP,FVP
651          COMMON/TST/PCRIT,A,B
652          COMMON/TMP/TC,TFR
653          LE10=2.302585093
654          IF(IU.LT.1)GOTO100
655          PSA=PSA*14.5
656      100 CONTINUE
657   C
658   C   CHECK PRESSURE.LE.CRITICAL PRESSURE
659   C
660          IF(PSA.GT.PCRIT) GO TO 902
661   C
662   C   COMPUTE INITIAL ESTIMATE OF 'TSAT'
663   C
664          PLOG=ALOG10(PSA)
```

140

```
665          TR=A*PLOG+B
666  C
667  C   ITERATE TO WITHIN .0001 DEG F   BY NEWTON ITERATION
668  C
669          DO 10 L=1,30
670          TR0=TR
671          C=ALOG10(ABS(FVP-TR0))
672          F=AVP+BVP/TR0+CVP*ALOG10(TR0)+DVP*TR0
673         1+EVP*((FVP-TR0)/TR0)*C-PLOG
674          FP=-BVP/TR0**2+CVP/(LE10*TR0)+DVP-EVP
675         1*(1./(LE10*TR0)+FVP*C/TR0**2)
676          TR=TR0-F/FP
677          IF(ABS(TR-TR0).LE.0.0001) GO TO 20
678      10 CONTINUE
679          WRITE(2,1000)
680      20 TSAT=TR-TFR
681          IF(IU.LT.1)GOTO200
682          PSA=PSA/14.5
683          TSAT=(TSAT-32.0)/1.8
684     200 RETURN
685     902 WRITE(2,1002)
686     999 PAUSE 'TSAT'
687    1000 FORMAT(/' *** WARNING *** TSAT : NOT CONVERGED'/)
688    1002 FORMAT(/' ***  ERROR  *** TSAT : PSA.GT.CRIT PRESSURE'/)
689          END


690          FUNCTION PSAT(NR,TF,IU)
691  C
692  C   DETERMINES THE SATURATION PRESSURE OF A REFRIGERANT
693  C   GIVEN TEMPERATURE
694  C
695  C          TF  = TEMPERATURE         DEG C (DEG F)
696  C          PSAT= ABSOLUTE PRESSURE  BAR   (P.S.I.A.)
697  C
698  C          IU = 0 : B.TH. UNITS   IU = 1 : S.I. UNITS
699  C
700          COMMON/SAT/AVP,BVP,CVP,DVP,EVP,FVP
701          COMMON/TMP/TC,TFR
702          IF(IU.LT.1)GOTO100
703          TF=TF*1.8+32.0
704     100 CONTINUE
705  C
706  C   CONVERT 'TF' TO ABSOLUTE TEMP (DEG R) AND CHECK IF ABOVE ZERO
707  C
708          T=TF+TFR
709          IF(T.LE.0.0) GO TO 902
710  C
711  C   CHECK IF TEMP.LE.CRITICAL TEMP
712  C
713          IF(T.GT.TC) GO TO 903
714  C
715      11 PSAT=10.**(AVP+BVP/T+CVP*ALOG10(T)+DVP*T+EVP*
716         1((FVP-T)/T)*ALOG10(ABS(FVP-T)))
717          IF(IU.LT.1)GOTO200
718          TF=(TF-32.0)/1.8
719          PSAT=PSAT/14.5
720     200 RETURN
721     902 WRITE(2,1002)
722          GOTO999
723     903 WRITE(2,1003)
724     999 PAUSE 'PSAT'
725    1002 FORMAT(/' ***  ERROR  *** PSAT : TEMP.LE.ZERO'/)
726    1003 FORMAT(/' ***  ERROR  *** PSAT : TEMP .GT.CRIT TEMP'/)
727          END


728          SUBROUTINE CKMSAT(NR,T,CPL,CPV,XKL,XKV,XMUL,XMUV)
729  C
730  C   DETERMINES SPECIFIC HEAT AT CONSTANT PRESSURE, THERMAL CONDUCTIVITY,
731  C   AND DYNAMIC VISCOSITY OF REFRIGERANTS AT SATURATION
732  C
733  C          *** S.I. UNITS ***
734  C
735          COMMON/CKMS/ACPL(7),ACPV(7),AKL(7),AKV(7),AML(7),
736         *AMV(7),TMINL,TMAXL,TMINV
737          IF(T.LT.TMINL.OR.T.GT.TMAXL)GOTO10
738          IF(T.LT.TMINV)WRITE(2,1002)
739          GOTO12
```

141

```
740      10 WRITE(2,1001)
741      12 CONTINUE
742         TT=1.0
743         CPL=0.0
744         CPV=0.0
745         XKL=0.0
746         XKV=0.0
747         XMUL=0.0
748         XMUV=0.0
749         DO 20 J=1,7
750         CPL=CPL+ACPL(J)*TT
751         CPV=CPV+ACPV(J)*TT
752         XKL=XKL+AKL(J)*TT
753         XKV=XKV+AKV(J)*TT
754         XMUL=XMUL+AML(J)*TT
755         XMUV=XMUV+AMV(J)*TT
756      20 TT=TT*T
757         RETURN
758    1001 FORMAT(/' *** WARNING *** CKMSAT : T OUT OF RANGE'/)
759    1002 FORMAT(/' *** WARNING *** CKMSAT : T OUT OF RANGE FOR VAPOUR ',
760        *'PROPERTIES'/)
761         END


762         SUBROUTINE CKMH2O(T,CP,XK,XMU)
763    C
764    C  DETERMINES SPECIFIC HEAT, CONDUCTIVITY, AND DYNAMIC VISCOSITY
765    C  OF WATER AT TEMPERATURE 'T'
766    C
767         IF(T.LT.0.0.OR.T.GT.300.0)WRITE(2,100)T
768         T2=T*T
769         T3=T2*T
770         T4=T3*T
771         T5=T4*T
772         T6=T5*T
773         CP=4.212238-2.475186E-03*T+6.298253E-05*T2-7.222080E-07*T3
774        *+4.786816E-09*T4-1.506013E-11*T5+1.896897E-14*T6
775         XK=5.689151E-04+1.874259E-06*T-7.888762E-09*T2-6.458946E-13*T3
776        *+5.693944E-14*T4-1.981186E-16*T5+2.158718E-19*T6
777         XMU=1.726597E-03-4.775260E-05*T+7.185304E-07*T2-6.070964E-09*T3
778        *+2.834833E-11*T4-6.790947E-14*T5+6.494669E-17*T6
779         RETURN
780     100 FORMAT(/' *** WARNING *** CKMH2O : TEMPERATURE OUT OF RANGE',
781        *F7.3,' C'/)
782         END
783         FINISH
```

```
1       REFRIGERANT DATA TO BE USED IN PROPERTIES PROGRAM SUITE.
2       (NOTE, THE CONSECUTIVE INTEGERS PRECEDING EACH DATA ELEMENT
3       ARE NOT PART OF THE DATA, BUT INDICATE THE ORDER OF ITEMS
4       IN THE DATA FILE)
5
6  11                          64     0.71750530000E-09     122     0.53341187000E 02
7     0.78117000000E-01        65    -0.28032670000E-11     123     0.00000000000E 00
8     0.19000000000E-02        66    -0.12786460000E-14     124     0.18691370000E 02
9    -0.31267590000E 01        67     0.94204530000E-04     125     0.00000000000E 00
10    0.13185230000E-02        68    -0.26500600000E-06     126     0.21983960000E 02
11   -0.35769990000E 02        69     0.40531290000E-09     127    -0.31509940000E 01
12   -0.25341000000E-01        70    -0.92329640000E-11     128     0.39883817270E 02
13    0.48751210000E-04        71    -0.42833390000E-13     129    -0.34366322280E 04
14    0.12203670000E 01        72     0.12269870000E-14     130    -0.12471522280E 02
15    0.16872770000E-02        73    -0.44708480000E-17     131     0.47304424400E-02
16   -0.18050620000E-05        74     0.21550930000E-04     132     0.00000000000E 00
17    0.00000000000E 00        75    -0.12338500000E-05     133     0.70000000000E 03
18   -0.23589300000E-04        76     0.41158650000E-07     134     0.80945000000E-02
19    0.24483030000E-07        77    -0.64519610000E-09     135     0.33266200000E-03
20   -0.14783790000E-03        78     0.52702970000E-11     136    -0.24138960000E-06
21    0.10575040000E 09        79    -0.21504410000E-13     137     0.67236300000E-10
22   -0.94721030000E 05        80     0.34575550000E-16     138     0.00000000000E 00
23    0.00000000000E 00        81     0.53647330000E-03     139     0.00000000000E 00
24    0.45000000000E 01        82    -0.59090770000E-05     140     0.39556551336E 02
25    0.58000000000E 03        83     0.55294780000E-07     141    -0.16537936091E-01
26    0.00000000000E 00        84    -0.53553550000E-09     142     0.59690000000E 03
27    0.84807000000E 03        85     0.44792560000E-11     143     0.12000000000E 03
28    0.45967000000E 03        86    -0.21181340000E-13     144     0.31200000000E 03
29    0.34570000000E 02        87     0.36941320000E-16     145     0.93221660000E 00
30    0.57638110000E 02        88     0.17790250000E-04     146     0.13067920000E-02
31    0.43632200000E 02        89    -0.88616730000E-06     147     0.70402160000E-05
32   -0.42823560000E 02        90     0.37335610000E-07     148     0.18441720000E-06
33    0.36706670000E 02        91    -0.68567670000E-09     149     0.28283690000E-08
34    0.00000000000E 00        92     0.62615270000E-11     150     0.78028820000E-11
35    0.00000000000E 00        93    -0.27693110000E-13     151    -0.68258120000E-13
36    0.42147028650E 02        94     0.47340360000E-16     152     0.64449560000E 00
37   -0.43443438070E 04        95    -0.73330000000E 02     153     0.26888840000E-02
38   -0.12845967530E 02        96     0.19778000000E 03     154     0.22648980000E-04
39    0.40083725000E-03        97     0.15560000000E 02     155    -0.30803670000E-07
40    0.31360535600E-01        98 12                        156    -0.43950340000E-08
41    0.86207000000E 03        99     0.88734000000E-01     157     0.63957500000E-10
42    0.23815000000E-01        100    0.65093886000E-02     158     0.50372580000E-12
43    0.27988230000E-03        101   -0.34097271300E 01     159     0.78698240000E-04
44   -0.21237340000E-06        102    0.15943484800E-02     160    -0.37454830000E-06
45    0.59990180000E-10        103   -0.56762767100E 02     161    -0.74090300000E-09
46    0.00000000000E 00        104    0.60239446500E-01     162     0.46133920000E-11
47   -0.33680703000E 03        105   -0.18796184300E-04     163     0.24327240000E-12
48    0.50541861087E 02        106    0.13113799800E 01     164    -0.51405640000E-15
49   -0.91839406797E-01        107   -0.54873701000E-03     165    -0.19696080000E-16
50    0.63950000000E 03        108    0.00000000000E 00     166     0.85773920000E-05
51    0.12480000000E 03        109    0.00000000000E 00     167     0.59869730000E-07
52    0.40870000000E 03        110    0.00000000000E 00     168    -0.15936340000E-08
53    0.86668240000E 00        111    0.34688340000E-08     169     0.50342170000E-11
54    0.74823940000E 03        112   -0.25439067800E-04     170     0.14552090000E-11
55    0.15367420000E-05        113    0.00000000000E 00     171    -0.26068320000E-13
56    0.22400430000E-07        114    0.00000000000E 00     172     0.12635120000E-15
57   -0.25152720000E-10        115    0.00000000000E 00     173     0.26470680000E-03
58   -0.20853770000E-11        116    0.54750000000E 01     174    -0.24522910000E-05
59    0.11324650000E-13        117    0.00000000000E 00     175     0.17536050000E-07
60    0.56017930000E 02        118    0.00000000000E 00     176    -0.10821490000E-09
61    0.12197090000E-02        119    0.69330000000E 03     177     0.15703240000E-11
62   -0.11696690000E-05        120    0.45970000000E 03     178    -0.15211140000E-13
63    0.14213520000E-07        121    0.34840000000E 02     179     0.24212790000E-17


180    0.12080300000E-04       243    0.18834260000E-12     306     0.54634409000E 02
181    0.55928490000E-07       244    0.31291840000E 00     307     0.36748920000E 02
182   -0.30511990000E-08       245    0.59635360000E-03     308    -0.22292565700E 02
183    0.12754770000E-10       246    0.10950320000E-05     309     0.20473288600E 02
184    0.24692060000E-11       247    0.57608730000E-07     310     0.00000000000E 00
185   -0.43296210000E-13       248   -0.79718890000E-11     311     0.00000000000E 00
186    0.20549740000E-15       249   -0.33440250000E-11     312     0.29357544530E 02
187   -0.95560000000E 02       250    0.23027520000E-13     313    -0.38451931520E 04
188    0.11222000000E 02       251    0.11182310000E-03     314    -0.78610312200E 02
189   -0.28890000000E 02       252   -0.35631220000E-06     315     0.21909390000E-02
190 21                         253    0.49169080000E-10     316     0.44574670300E 00
191    0.10427000000E 00       254   -0.12275590000E-10     317     0.68610000000E 03
192    0.00000000000E 00       255    0.29359540000E-13     318     0.28128360000E-01
193   -0.73160000000E 01       256    0.12124010000E-14     319     0.22554080000E-03
```

```
194   0.46421000000E-02     257  -0.61729280000E-17     320  -0.65096070000E-07
195   0.00000000000E 00     258   0.12210810000E-04     321   0.00000000000E 00
196  -0.20382376000E 00     259  -0.53188200000E-06     322   0.00000000000E 00
197   0.35930000000E-03     260   0.25227650000E-07     323   0.25734100000E 03
198   0.00000000000E 00     261  -0.50941700000E-09     324   0.62401097315E 02
199   0.00000000000E 00     262   0.51772170000E-11     325  -0.45333083284E-01
200   0.00000000000E 00     263  -0.25697580000E-13     326   0.72190600000E 03
201   0.00000000000E 00     264   0.49547780000E-16     327   0.12000000000E 03
202   0.00000000000E 00     265   0.39337870000E-03     328   0.38800000000E 03
203   0.00000000000E 00     266  -0.37132620000E-05     329   0.11731540000E 01
204   0.00000000000E 00     267   0.30398350000E-07     330   0.23384480000E-02
205   0.00000000000E 00     268  -0.30216660000E-09     331   0.16943970000E-04
206   0.00000000000E 00     269   0.27108890000E-11     332   0.29298850000E-06
207   0.00000000000E 00     270  -0.11970670000E-13     333   0.45565530000E-08
208   0.00000000000E 00     271   0.14868680000E-16     334   0.42416000000E-11
209   0.00000000000E 00     272   0.19176190000E-04     335  -0.18710980000E-12
210   0.00000000000E 00     273  -0.10514960000E-05     336   0.71421990000E 00
211   0.81290000000E 03     274   0.46992270000E-07     337   0.36432310000E-02
212   0.45969000000E 03     275  -0.92767020000E-09     338   0.47881710000E-04
213   0.11637962000E 03     276   0.91646830000E-11     339   0.49450150000E-06
214  -0.31068080000E-01     277  -0.44013530000E-13     340  -0.13997060000E-08
215  -0.50100000000E-04     278   0.81889650000E-16     341  -0.77823860000E-14
216   0.00000000000E 00     279  -0.85000000000E 02     342   0.25244620000E-12
217   0.00000000000E 00     280   0.17890000000E 03     343   0.10060710000E-03
218   0.00000000000E 00     281   0.15000000000E 02     344  -0.49451370000E-06
219   0.00000000000E 00     282  22                     345  -0.15154320000E-03
220   0.42790800000E 02     283   0.12409800000E 00     346  -0.14274060000E-11
221  -0.42613400000E 04     284   0.20000000000E-02     347   0.72786410000E-12
222  -0.13029500000E 02     285  -0.43535470000E 01     348   0.18051460000E-15
223   0.39851000000E-02     286   0.24072520000E 00     349  -0.87262210000E-16
224   0.00000000000E 00     287  -0.44066868000E 02     350   0.95049310000E-05
225   0.00000000000E 00     288  -0.17464000000E-01     351   0.43605070000E-07
226   0.42700000000E-01     289   0.76278900000E-04     352  -0.44421610000E-09
227   0.14000000000E-03     290   0.14837630000E 01     353   0.37665340000E-10
228   0.00000000000E 00     291   0.23101420000E-02     354   0.38476540000E-13
229   0.00000000000E 00     292  -0.36057230000E-05     355  -0.16128920000E-13
230   0.00000000000E 00     293   0.00000000000E 00     356   0.14651150000E-15
231   0.00000000000E 00     294  -0.37240440000E-04     357   0.23471150000E-03
232   0.762470363713E 02    295   0.53554650000E-07     358  -0.16445890000E-05
233  -0.110614315265E 00    296  -0.18450510000E-03     359   0.68228320000E-08
234   0.74940000000E 03     297   0.13633870000E 09     360  -0.97040090000E-10
235   0.10677000000E 03     298  -0.16726120000E 06     361   0.28747120000E-11
236   0.38851000000E 03     299   0.00000000000E 00     362  -0.35030320000E-14
237   0.56316690000E 00     300   0.42000000000E 01     363  -0.29042920000E-15
238   0.74843590000E 00     301   0.54820000000E 03     364   0.12017870000E-04
239   0.11099250000E-04     302   0.00000000000E 00     365   0.11666360000E-07
240   0.12722410000E-07     303   0.66450000000E 03     366  -0.79335080000E-09
241  -0.16287510000E-08     304   0.45969000000E 03     367   0.76673640000E-10
242  -0.11747090000E-11     305   0.32760000000E 02     368   0.19245380000E-12


369  -0.31579830000E-13     432   0.25651820000E-07     495   0.00000000000E 00
370   0.25127750000E-15     433  -0.30487250000E-09     496   0.10644955000E 02
371  -0.73330050000E 02     434   0.11800550000E-11     497  -0.36711538130E 04
372   0.96110000000E 02     435   0.71057270000E-04     498  -0.36983500000E 00
373  -0.40000000000E 02     436  -0.26044710000E-06     499  -0.17463520000E-02
374  114                    437  -0.58441830000E-12     500   0.81611390000E 00
375   0.62780807000E-01     438  -0.20907460000E-12     501   0.65400000000E 03
376   0.59149070000E-02     439   0.63779520000E-14     502   0.20419000000E-01
377  -0.23856704000E 00     440   0.92245940000E-17     503   0.29968020000E-03
378   0.10801207000E-02     441  -0.14558740000E-17     504  -0.14090430000E-06
379  -0.65643648000E 01     442   0.88955400001E-05     505   0.22100610000E-10
380   0.34055687000E 00     443   0.73681240000E-07     506   0.00000000000E 00
381  -0.53336494000E-05     444  -0.95209330001E-09     507   0.00000000000E 00
382   0.16366057000E 00     445   0.23577710000E-10     508   0.35308000000E 02
383  -0.38574810000E 00     446  -0.28705460000E-12     509  -0.74440000000E-01
384   0.00000000000E 00     447   0.16859640000E-14     510   0.59100000000E 03
385   0.00000000000E 00     448  -0.37630800000E-17     511   0.11700000000E 03
386   0.16017659000E-05     449   0.46410490000E-03     512   0.27900000000E 03
387   0.62632341000E-09     450  -0.61407620000E-05     513   0.65474260000E 00
388  -0.10165314000E-04     451   0.59717580000E-07     514   0.15350000000E-02
389   0.00000000000E 00     452  -0.64132020000E-09     515   0.51981540000E-05
390   0.00000000000E 00     453   0.70820510000E-11     516  -0.43372110000E-07
391   0.00000000000E 00     454  -0.42697020000E-13     517  -0.11502360000E-08
392   0.30000000000E 01     455   0.83521030000E-16     518   0.72883030000E-11
393   0.00000000000E 00     456   0.98618870000E-05     519   0.18407550000E-12
394   0.00000000000E 00     457   0.23509520000E-07     520   0.38930520000E 00
395   0.75395000000E 03     458   0.10236880000E-07     521   0.14513460000E-02
396   0.45969000000E 03     459  -0.46545670000E-09     522   0.33191540000E-04
```

144

397	0.36320000000E 02	460	0.78915530000E-11	523	0.31060100000E-06
398	0.61146414000E 02	461	-0.57357390000E-13	524	-0.81853410001E-08
399	0.00000000000E 00	462	0.15130660000E-15	525	0.70957730000E-10
400	0.16418015000E 02	463	-0.73300000000E 02	526	0.23270830000E-11
401	0.00000000000E 00	464	0.14560000000E 03	527	0.74587130000E-04
402	0.17476838000E 02	465	0.44000000000E 01	528	-0.39981220000E-06
403	0.11198280000E 01	466	502	529	-0.13365500000E-08
404	0.27071306000E 02	467	0.96125000000E-01	530	0.95085410000E-11
405	-0.51137021000E 04	468	0.16700000000E-02	531	0.70440820000E-12
406	-0.63086761000E 01	469	-0.32613344000E 01	532	-0.30339740000E-14
407	0.69130030000E-03	470	0.20576287000E-02	533	-0.10891310000E-15
408	0.78142111000E 00	471	-0.24248790000E 02	534	0.10175750000E-04
409	0.76835000000E 03	472	0.34866748000E-01	535	0.25803380000E-07
410	0.17500000000E-01	473	-0.86791313000E-05	536	0.15256200000E-09
411	0.34900000000E 03	474	0.33274779000E 00	537	0.63977240000E-10
412	-0.16700000000E-06	475	-0.85765677000E-03	538	-0.50576560000E-12
413	0.00000000000E 00	476	0.70240549000E-06	539	-0.26654790000E-13
414	0.00000000000E 00	477	0.22412368000E-01	540	0.31826830000E-15
415	0.00000000000E 00	478	0.88368966999E-05	541	0.23347680000E-03
416	0.25339662100E 02	479	-0.79168095000E-08	542	-0.22994650000E-05
417	-0.11513713000E 00	480	-0.37167231000E-03	543	-0.72631470000E-09
418	0.47000000000E 03	481	-0.38257766000E 08	544	-0.95180680000E-10
419	0.13360000000E 03	482	0.55816094000E 05	545	0.75782510000E-11
420	-0.11490000000E 03	483	0.15378377000E 10	546	-0.14742360000E-13
421	0.95356170000E 00	484	0.42000000000E 01	547	-0.77996010000E-15
422	0.23076560000E-02	485	0.60900000000E 03	548	0.11574040000E-04
423	0.39871160000E-06	486	0.70000000000E-06	549	0.27298330000E-07
424	-0.15931440000E-08	487	0.63956000000E 03	550	0.17361310000E-08
425	0.11696800000E-09	488	0.45967000000E 03	551	0.43476100000E-10
426	0.89170470000E-12	489	0.35000000000E 02	552	-0.12766180000E-11
427	-0.11637750000E-13	490	0.53484370000E 02	553	-0.12390020000E-13
428	0.68773550000E 00	491	0.63864170000E 02	554	0.27643580000E-15
429	0.22552010000E 02	492	-0.70080660000E 02	555	-0.74000000000E 02
430	-0.23844610000E-04	493	0.48479010000E 02	556	0.82200000000E 02
431	-0.42643370000E-06	494	0.00000000000E 00	557	-0.52000000000E 02

```
SAMPLE OUTPUT FROM PROGRAM SUITE.
INPUT DATA IS SHOWN AS <NNN
UNITS ARE OMITTED FROM NORMAL OUTPUT




     REFRIGERANT PROPERTIES PROGRAM
     ********************************

REFRIGERANT NO ?
<  12

UNITS ?  TYPE 0 FOR B.TH.UNITS
              1 FOR S.I. UNITS
<  1


PRESSURE ? (= 0 TO STOP PROGRAM)
<  3.0

SECOND PARAMETER ?  TYPE 1 FOR TEMPERATURE
                         2 FOR SPECIFIC VOLUME
                         3 FOR SPECIFIC ENTHALPY
                         4 FOR SPECIFIC ENTROPY
                         5 FOR QUALITY

<  1

TEMPERATURE ?
<15.0


R 12

T =   15.00      (DEG C)
P =    3.00000   (BAR)
V =    0.06138   (M.CUB PER KG)
H =  197.40405   (KJ PER M CUB)
S =    0.73348   (KJ PER M CUB PER DEG K)
X =    1.06743   (QUALITY)



PRESSURE ? (= 0 TO STOP PROGRAM)
<0
12.30.25←
```

REFERENCES

Alfred *c*900 *The Voyage of Wulfstan*

ASHRAE 1973 *Systems Guide* (New York: ASHRAE)

—— 1974 *Handbook of Fundamentals* (New York: ASHRAE)

Bambach G 1955 *Das Verhalten von Mineralöl — F12 — Gemischen in Kaltemaschinen* (Karlsruhe: C F Muller)

Blundell C J 1978 Efficient dehumidification *Heating and Ventilating Engineer* September 5–7

Bredesen A M 1979 Effects of speed control on the valve performance and on the energy efficiency of heat pump reciprocating compressors *Antriebe für Wärmepumpen* ed. J Paul (Essen: Vulkan-Verlag)

Buick T R, McMullan J T, Morgan R and Murray R B 1978 Ice detection in heat pumps and coolers *Int. J. Energy Research* **2** (1) 85–98

Central Policy Review Staff 1974 *Energy Conservation* (London: HMSO)

CIB 1977 Performance specification for domestic heat pumps *CIB Report* W67

CIBS 1980 *CIBS Guide* (London: Chartered Institute of Building Services)

Cooper K W and Mount A G 1972 Oil circulation — its effect on compressor capacity, theory and experiment *Purdue Compressor Technology Conference, Purdue University* (Purdue Research Foundation)

Downing R C 1972 Refrigerant equations *ASHRAE Trans.* **78** 158–69

Dutram L L and Sarkes L A 1979 Natural gas heat pump implementations and development *Antriebe für Wärmepumpen* ed. J Paul (Essen: Vulkan-Verlag) pp 86–92

EEC 1978 Guidelines for the presentation of final reports *Commission of the European Communities Report* XII/962-1/78-EN

Ek A 1978 Variable-speed a.c. drives for severe environments *ASEA J.* **51** (2) 35–40

Fanger P O 1970 *Thermal Comfort* (New York: McGraw-Hill)

Farrell T 1975 Thermoelectric heat pumping *Electricity Council Research Centre Report* ECRC/R844

Field A A 1976 *Heating and Ventilating Engineer* October 6–10

Fleming H 1978 A variable-speed heat pump *MPhil Thesis* Manchester Polytechnic

Heap R D 1977 American heat pumps in British houses *Elektrowärme in Technischen Ausbau* **35** (A2) 77–81

van Heyden L and Wölting W 1979 Übersicht der projektieren Gaswärmepumpen —— Anlagen grösserer Leistung mit stationären Motoren *Antriebe für Wärme pumpen* ed. J Paul (Essen: Vulkan-Verlag) pp 124–7

Hiller C C and Glicksman L R 1976 Improving heat pump performance via compressor capacity control *MIT Energy Laboratory Report* MIT–EL76–001

Hodgett D L and Lincoln P 1978 A mathematical model of a dehumidifying evaporator for high temperature heat pumps *Electricity Council Research Centre Report* ECRC/M1147

Hughes D, McMullan J T, Morgan R and Mawhinney K A 1980 Influence of lubricant on heat pump performance *Purdue Compressor Technology Conf., Purdue University* (Purdue Research Foundation)

de Jong H C J 1976 *A.C. Motor Design* (Oxford: Oxford University Press)

Kartsounes G T and Erth R A 1971 Computer calculations of the thermodynamic properties of refrigerants 12, 22 and 502 *ASHRAE Trans.* **77** 88–103

McMullan J T and Morgan R 1979 The development of domestic heat pumps *Mtg. on Heat Pump Research, Development and Application* ed. H Ehringer and G Hoyaux (Commission of the European Communities Publication EUR 6237)
—— 1981 to be published

McMullan J T, Morgan R and Hughes D 1980 Development of domestic heat pumps *New Ways of Saving Energy* ed. A Strub and H Ehringer (Dordrecht, Netherlands: D Reidel)

McMullan J T, Morgan R and Murray R B 1976 *Energy Resources and Supply* (Chichester: Wiley)
—— 1977 A controlled environment laboratory for the testing of domestic heat pumps *Int. J. Energy Res.* **1** 47–54

Martin J J and Hou Y C 1955 Development of an equation of state for gases *AIChE J.* **1** 142

Morgan R and McMullan J T 1980 Operating characteristics of air to water heat pumps in Belfast during the winters of 1978/9 and 1979/80, *C.I.B. W.67 technical subgroup meeting, Copenhagen*

Muirchu 7th Century *Life of St. Patrick* vol 1 Ch 19

Nevins R G, Michaels K B and Feyerherm A M 1964 Effect of floor surface temperature on comfort Part 1 College age males *ASHRAE Trans.* **70** 29 Part 2 College age females *ASHRAE Trans.* **70** 37

Nevins R G, Rohles F H, Springer W and Feyerherm A M 1966 A temperature–humidity chart for thermal comfort of seated persons *ASHRAE Trans.* **72** 283–91

Pascente J 1979 Energy-saving motor control *New Electronics* **12** (7) 79

Paul J 1979 Betriebseigenschaften von Kolbenverdichtern für Wärmepumpen bei veranderlichen Drehzahl unter Berüchsichtigung des Ventilverhalterns *Antriebe für Wärmepumpen* ed. J Paul (Essen: Vulkan-Verlag)

Paul J and Steimle F 1980 Experiences with air to water and water to water heat pumps operating in existing buildings — ways of improving the cop of heat pumps for hydronic heating systems *Int. Congr. on Building Energy Management, Portugal*

Pegley A C and Rieke A 1979 Small gas engines as prime movers for heat pumps in domestic heating *Antriebe für Wärmepumpen* ed. J Paul (Essen: Vulkan-Verlag) pp 41–6

Pollard R 1975 Heat pumps — prospects and problems for domestic units *Electrical Review* **3** 425–7

Rayment R and Morgan K 1980 The performance of heating control systems in experimental houses *New Ways of Saving Energy* ed. A Strub and H Ehringer (Dordrecht, Netherlands: D Riedel)

Reay D A 1979 *Heat Recovery Systems* (London: E and F N Spon)

Richardson D W and Husker C 1975 An ice-free evaporator *GB Patent Specification* 1463 210

Rojey A, Meyer C, Choffe B, Jacq J, Asselineau L and Vidal J 1980 Heat pump with a fluid mixture *New Ways of Saving Energy* ed. A Strub and H Ehringer (Dordrecht, Netherlands: D Reidel) pp 242–51

Rummel Th 1979 Wärmepumpen mit Dieselmotorantrieb und mit kombinierten Dieselmotor–Elektromotor–Antrieb *Antrieb für Warmepumpen* ed. J Paul (Essen: Vulkan-Verlag) pp 107–10

Strong D T G 1979 Directly-fired domestic heat pump *Antriebe für Wärmepumpen* ed. J Paul (Essen: Vulkan-Verlag) pp 95–102

Trask A 1979 Heat pumps — the defrost problem *ASHRAE J.* **21** (9) 33–6

Ubbels J, Meulman A P and Verheij C P 1980 The saving of energy when cooling milk and heating water on farms *New Ways of Saving Energy* ed. A Strub and H Ehringer (Dordrecht, Netherlands: D Reidel)

Winkler J P and Schneider P 1979 Wärmekraftkopplingsanlagen—Kriterien der Dimensionierung *Antriebe für Wärmepumpen* ed. J Paul (Essen: Vulkan-Verlag) pp 111–23

BIBLIOGRAPHY

Chapter 1

Ambrose E R 1966 *Heat Pumps and Electric Heating* (New York: Wiley)

BCC 1978 *Heat Pumps: Commercial Survey* (Stamford: Business Communications Co.)

Berridge G L C 1975 *Heat Pumps — Key References in the Literature* (Boston Spa: Brainchild Information Services)

Collie M J 1979 *Heat Pump Technology for Energy Saving* (Park Ridge, NJ: Noyes Data Corp.)

Davies S J 1950 *Heat Pumps and Thermal Compressors* (London: Constable)

EEC 1978 Study day on the development of heat pumps in the community for heating and air-conditioning *Commission of the European Communities Report* EUR 6161

Ehringer H and Hoyaux G (eds) 1979 *Mtg on Heat Pumps Research, Development and Application* (Commission of the European Communities Publication EUR 6237)

Ellington R T, Knust G, Peck R E and Read J F 1957 The absorption cooling process *Institute of Gas Technology Research Bulletin No* 14

Foster J 1979 A review of heat pump research in UK universities, polytechnics and institutes of technology *Science Research Council Report* RL–79–050

Haldane T G N 1930 The heat pump — an economical method of producing low grade heat from electricity *J. IEEE* 666–75

Heap R D 1979 *Heat Pumps* (London: E and F N Spon)

Heating, Ventilating, Refrigeration and Air-Conditioning Year Book 1978 (London: Heating and Ventilating Contractors' Association)

Kemler E N and Oglesby S 1950 *Heat Pump Applications* (New York: McGraw-Hill)

Lloyd S and Starling C 1975 Heat pumps *BSRIA Bibliography* 103

Macmichael D B A and Reay D A 1979 *Heat Pumps, Theory, Design and Applications* (Oxford: Pergamon Press)

Sumner J A 1975 *Domestic Heat Pumps* (Dorchester: Prism Press)

——*Introduction to Heat Pumps* (Dorchester: Prism Press)

Chapter 2

Ambrose E R 1974 The heat pump: performance factor, possible improvements *Heating, Piping and Air-conditioning* May 77–82

Blundell C J 1977 A flat plate evaporator for domestic heat pumps *Electricity Council Research Centre Report* ECRC/N1077

150

——1977 Optimising heat exchangers for air-to-air space heating heat pumps in the UK *Int. J. Energy Res.* **1** 69–94

Colburn A P and Hougen O A 1934 Design of cooler condensers for mixture of vapours with non-condensing gas *Ind. Engng. Chem.* **26** 1178–82

Dossat R J 1978 *Principles of Refrigeration* (New York: Wiley)

Farrell T 1975 Thermoelectric heat pumping *Electricity Council Research Centre Report* ECRC/R844

Kays W M and London A L 1954 *Compact Heat Exchangers* (New York: McGraw-Hill)

Kruse H and Jakobs R 1977 The importance of non-azeotropic binary refrigerants for use in heat pumps and refrigerating plant *Klima and Kalte Ing.* July/Aug 253–60

McElgin J and Wiley D C 1940 Calculation of coil surface areas for air cooling and dehumidification *Heating, Piping and Air-conditioning* March 195–201

McQuiston F C 1976 Heat mass and momentum transfer in a parallel plate dehumidifying heat exchanger *ASHRAE Trans.* **82** 87–106

Midgley T and Henne A L 1930 Organic fluorides as refrigerants *Ind. Engng. Chem.* **22** 542

Chapter 3

Edwards J D 1973 *Electrical Machines* (Aylesbury: International Textbook Co.)

Matsch L W 1972 *Electromagnetic and Electromechanical Machines* (Scranton, Pennsylvania: Intext Educational Publishers)

Chapter 4

ASHRAE 1975 *Equipment Handbook* (New York: ASHRAE)

Heap R D 1979 *Heat Pumps* (London: E and F N Spon)

Chapter 5

Braham G D and Johnson A J 1977 *A Guide to Energy Cost Effectiveness in Swimming Pools* (London: Electricity Council)

Camantini E and Kestler T (eds) 1976 *Heat Pumps and Their Contribution to Energy Conservation* (Leyden, Netherlands: Noordhoff)

Courtney R G 1976 *Energy Conservation in the Built Environment* (Lancaster: Construction Press/CIB)

Freund P, Leach S and Seymor-Walker K 1976 Heat pumps for use in buildings *Building Research Establishment Current Paper* 1976

Geeraert B 1976 Air drying by heat pumps, with special reference to timber drying *Heat Pumps and Their Contribution to Energy Conservation* ed. E Camantini and T Kestler (Leyden, Netherlands: Noordhoff) pp219–46

Heap R D 1979 *Heat Pumps* (London: E and F N Spon Ltd)

Hodgett D L 1976 Improving the efficiency of drying using heat pumps *Electricity Council Research Centre Report* ECRC/M956

Janich H 1978 Experience in the design, installation and operation of heat pumps using ground water as source of heat *Elektrowarme Int.* **36** (AR) A107–110 (in German)

Kemler E N and Oglesby S 1950 *Heat Pump Applications* (New York: McGraw-Hill)

Kernan G and Brady J 1977 Economic evaluation of heat pumps *Int. J. Energy Res.* **1** 115–25

Kolbusz P 1972 The improvement of drying efficiency *Electricity Council Research Centre Report* ECRC/R476

——1975 Industrial applications of heat pumps *Electricity Council Research Centre Report* ECRC/N845

Ledermann H 1977 Space heating with heat pump and thermal storage *Bull. Assoc. Suisse Elect.* **68** 185–7

Macmichael D B A and Reay D A 1979 *Heat Pumps, Theory, Design and Applications* (London: Pergamon)

Michel H, Pottier J and Jaéglé J 1977 Les Pompes à Chaleur à Compression et à Absorption — Domaines d'emploi et Perspectives de Developpement *Proc. 10th World Energy Conference, Istanbul* paper 4.6–1

Sherratt A F C 1974 *Integrated Environment in Building Design* (London: Applied Science Press)

Siviour J B 1977 Ranking energy saving ideas *Electricity Council Research Centre Report* ECRC/M1042

INDEX